T0230448

The Manual of Strategic Economic
Decision Making

Jeff Grover

The Manual of Strategic Economic Decision Making

Using Bayesian Belief Networks to Solve Complex Problems

 Springer

Jeff Grover
Grover Group, Inc.
Bowling Green, KY, USA

ISBN 978-3-319-83937-0   ISBN 978-3-319-48414-3   (eBook)
DOI 10.1007/978-3-319-48414-3

Printed on acid-free paper

This Springer imprint is published by Springer Nature
The registered company is Springer International Publishing AG
The registered company address is: Gewerbestrasse 11, 6330 Cham, Switzerland

*In memory of Stanley C. Brown
and Michael Brown*

Executive Summary

As I researched and published *Strategic Economic Decision-Making: Using Bayesian Belief Networks to Solve Complex Problems* (Grover 2013), it became clear that the direction of the application of Bayes' theorem[1] and Bayesian belief networks (BBN) is toward the computer algorithm community. To this end, we created our BBN algorithm, BayeSniffer™ 2014–2016, and coded in using the Microsoft Structured Query Language and ported it to www.BayeSniffer.com. Now, we have a fully functioning algorithm to parse through fully specified BBN. This motivates us to provide a grounded approach to the basic mathematical and statistical constructs of the theorem. This should assist the naïve learner's understanding of its application to the decision-making process across the spectrum of economic events. It is evident from the literature that industry has greatly benefited from the utility of this theorem. In fact, a current Google search using the keyword "Bayesian" populated more than twelve million hits as compared to only four million hits during the writing of the first manuscript. We predict exponential interest in the utility of the theorem and BBN in the next decade. We have governments and private sector industry portals, especially Twitter, Facebook, and Google, using high-performance computers to collect enormous amounts of data globally. These portals are collecting vast amounts of information based on human decision-making. Understanding this phenomenon could greatly assist economic decision-making across these economies. The goal of this manual is to provide the learner with simple constructs to use in converting this information into business intelligence for decision-making, and BBN are proving to be the tool of choice in this endeavor. A limitation to learning BBN is that the statistical and computer programming symbologies are not straightforward, which is creating a constraint for the learner. This motivates us to provide a manual as a prescriptive guide for building BBN to assist in decision-making processes. We have stripped out all statistical jargon to provide the BBN community and the learner with a concise manual for this purpose.

---

[1] Hereafter we will also refer to Bayes' theorem as "the theorem."

Acknowledgments

Some fathers and sons bond with hunting and fishing. A special thanks to Jeffrey. I would also like to thank the following contributors: Lydia Parks (editor/Copy), Philip B. Stark, Ph.D., Anna T. Cianciolo, Ph.D., Stefan Conrady, Major Tonya R. Tatum, and Denise Vaught, MBA.

Contents

List of Figures

List of Tables

Chapter 1
Introduction

1.1 Prologue

Please note, there are no discoveries in this manual. Others have paved the path, but the learnings are new for me, and thus the reader.

1.2 Quintessential Bayes'

The essence of Bayes' theorem is contained in the following thoughts:

- Things happen together.
- One thing happens, because another one has happened.
- Two things happening together, is the same, as the happening of the first one, given the second one happens.
- The happening of a first thing, given the second one happens, is the same, as the magnitude of either one of the two things happening given the happening, of the second thing, and the second thing happening.
- The happening of a first thing, given the second one happens, is also the same as the proportion of both things happening, together by the second thing happening.
- The happening of a first thing, given the second one happens, is not the same as the second one happening, given the happening of the first thing.

Based on the laws of probability and the propagating rule of multiplication, we can truthify the information contained in discrete or discretionized data, with the limitation of computing processing power. This manual provides a keystone approach to problem solving using selected joint and disjoint spaces, to harness the power of the theorem and Bayesian Belief Networks (BBN).

© Springer International Publishing AG 2016   1
J. Grover, *The Manual of Strategic Economic Decision Making*,
DOI 10.1007/978-3-319-48414-3_1

1.3   Scope

This manual serves as a reference showing how to use joint and disjoint BBN using discrete mathematics. Using this method, we can fully understand the fundamental principles of BBN, and apply them to decision-making processes. We will provide the learner with a set of base matrices for two, three and four node BBN to use in fitting this series of BBN. Using this protocol will allow them to quickly master those areas that require a more theoretical and abstract use of the theorem.[1]

1.4   Motivation

We are motivated to share the concept of the theorem due to its simplicity and utility for everyday life. It has an uncanny ability to separate truth from fiction, and can truthify information rigidly, logically, and quickly. As we experience life events, we can apply inductive logic and inverse probability to noisy information and point toward the truth. The global economy bombards us with vast amounts of information, from massive amounts of data. If we are searching for the truth, we can use the inductive logic of the theorem, as a filter, to see it clearly. We are also motivated to share this concept with strategic decision-makers since the theorem absorbs subject matter expertise (SME), without the need of data. Knowing what questions to ask will empower them to make the right decisions using all possible information.

1.5   Intent

My intent is to present the elementary principles of the theorem using minimal statistical terminology and symbology. As indicated earlier, the learning of this is difficult due to its complex nature, and scattered statistical symbology. In this manual, we will bridge this gap by providing grounded constructs. Having a fundamental understanding of these constructs is essential in absorbing the concepts embedded in the theorem. We will provide the learner with the basic statistical concepts, terminology, and definitions to illustrate the concepts put forth here. Decision-makers can also benefit from this logic, by providing their subjective SME in BBN to obtain hidden truths of their knowledge.

---

[1]The basis for the data consist of a 10,000 rows of randomly generated data from a Monte Carlo experiment to build up to and including a four node BBN.

1.6   Utility

The utility of the theorem reaches across all branches of science. While very simple in design, it requires sound, inductive logic when applied to causal relationships. This motivates us to combine past outcomes with observable events to illuminate the truth. What has great value in the Bayesian universe is its utility to contain an infinite number of illuminating events that when invoked, scale them down, so that the revised space begins to learn the truth. We begin with a subjective view of what we believe the truth is, and then by applying Bayesian theory, it begins to reveal itself. This is opposed to the philosophy of deductive, logic of falsification, which never really accepts the truth.

1.7   Introduction to Bayes' Theorem and Bayesian Belief Networks

The demand of determining future states of nature based on complex streams of asymmetric information is increasing and comes at a premium cost for today's organizations across a global economy. Strategic leaders at all levels face events where information feeds at near real-time, requiring decision-making based on interactive effects across all spectrums of operations. The dominant information that has historically been absent here is subjective in nature, and flows directly from the innate knowledge of leaders and SME of these organizations. With the use of inductive reasoning, we can integrate this knowledge, or degrees of truth, and have a more plausible future expectation of event outcomes based on experience by filtering it through the lens of the theorem. We do this by formulating a hypothesis (a cause) of proportional relationships that we believe exist (the cause), and then filtering it through observable information (the effect/s) to revise an initial belief.

There is a gradual acceptance by the scientific community of traditionalists (frequentists) for the Bayesian methodology. This is not by a new theoretical revelation, but through its current utility in scientific discovery. It allows researchers to seamlessly transition from the traditional cause and effect to the effect and cause scenario using plausible reasoning.[2] This philosophy is possible, in part, through the use of subjective (prior) beliefs where researchers obtain knowledge, either through historical information or from SME when attempting to formulate the truth. This knowledge[3] can either originate from observable data, or intuitive facts, as seen

---

[2]E. T. Jaynes, in his book, *Probability Theory: The Logic of Science* (Jaynes, 1995) suggests the concept of plausible reasoning is a limited form of deductive logic and "The theory of plausible reasoning" … "is not a weakened form of logic; it is an extension of logic with new content not present at all in conventional deductive logic" (p. 8).

[3]In the BBN literature, researchers refer to this knowledge as subjective and originates from *a priori* (prior) probabilities.

through the lenses of these SME. The use of this *prior* knowledge manifests a theoretical clash between the traditionalists and Bayesians. The question remains, "Does the Bayesian method have utility?" In response to this question, we can offer an example from the art and science of diagnosing a disease. There is a consensus that the science of diagnosing is exact. The art of diagnosing, however, is not exact. Let's consider a case and solve it using the basic concepts of the theorem. Suppose you undergo a medical test to rule out a disease and the test result is positive (event A), which suggests that you have the disease (event B). We are conditioning event A on B and expressing the relationship that says, the probability of the test being positive, $P(A)$, given you have the disease $P(B)$, or $P(A|B)$.

What we are looking for is the opposite—the probability that you have the disease, given the test is positive, $P(B|A)$, which becomes our hypothesis (medical test results). If we think the prevalent rate of people in the universe that actually have the disease is 1.0 %, $P(B)$, and let $P(A|B) = 95.0\%$ (our hypothesis), this suggests that 95 out of 100 people that test positive for the disease, actually have the disease, and 99 out of 100 people that tested negative actually do not have the disease, where $P(\tilde{A}|\tilde{B}) = 99.0\%$. Now, revealed is a clearer picture of the truth and we see that the chance of actually having has the disease is considerably less, $P(A) = 5.9\%$.[4] Now we have enough information to answer $P(B|A)$. Using the' theorem, Eq. (1.1), we can answer our question:

$$P(B|A) = \frac{P(A|B) * P(B)}{P(A)} \tag{1.1}$$

Now, we can simply solve this equation and obtain our answer in Eq. (1.2).

$$P(B|A) = \frac{(95.0\%) * (1.0\%)}{5.9\%} = \frac{0.95\%}{5.9\%} = 16.1\%. \tag{1.2}$$

The probability of actually having the disease given you have a positive test result, is downgraded from 95 to 16.1 %, which is a 490.1 %[5] reduction in the initial hypothesis, compared to just chance alone. This is the essence of the theorem— it has the ability to slice through observable information using prior beliefs to proportionally reweight the truth.

Now, consider the scenario where your physician makes a diagnosis based on the original test results of 95.0 % and recommends surgery, a regiment of medicine, or even additional tests. If she or he is incorrect in their diagnosis then the economic consequences at a minimum not only include the psychological costs of mental, physical and emotional pain and suffering, but also the costs associated with surgery and a regiment of medication. Often the cure is worse than the disease. When your

---

[4]We computed the (marginal) probability of event A using the law of total probability as: $P(A) = P(A, B) + P(A, B') = P(A|B)^* P(B) + P(A|B')^* P(B') = 95.0\%^* 1.0\% + 5.0\%^* 99.0\% = 0.95\% + 4.95\% = 5.9\%$.

[5]We computed this as: 490.1 % $= (95\% - 16.1\%)/16.1\%$.

physician begins to add prior knowledge or initial beliefs to his or her case, the original diagnosis comes into question. Since only 1.0 % of the population actually has this disease, P(B) = 1.0 %, using inductive logic, your physician would begin to adjust their initial beliefs of the diagnosis downward. What if your physician adds the fact that there is no family history? Then they would continue to adjust their beliefs, possibly to non-alarming levels.[6] This manifests the underlying principles of BBN; they learn from these partial truths or knowledge. Inductively, you would be less confident in the initial diagnosis, if your physician did not consider these initial truths when making their diagnosis.

1.8   Inductive Versus Deductive Logic

The frequentists argue that deductive reasoning is the only way to the truth, and the Bayesian argues that the past reveals the truth. Of course, while the former will immediately suggest that the latter is biasing their data selection process, by either using subjective reasoning or reaching back to historical or past observable events to determine future states of nature, the former remains in a theoretical rut by not illuminating the truth using this subjective information. While at the same time, the Bayesians are exponentially exploiting the universe of truth by doing this reach-back and suggesting reasonable future states of nature. Ask Twitter, Microsoft, or Google about their use of Bayesian inference within their search engines, or ask the medical community when they correctly diagnose the existence or nonexistence of cancer. Clearly, there is a utility in the theorem and the use of inductive logic.

In defining inductive and deductive logic, Bolstad (2007) suggests the former uses plausible reasoning, to infer the truth contained within a statement, to gauge the truth or falsehood of other statements that are consequences of the initial one. He also suggests that inductive logic goes from the specific to the general, using statistical inferences of a parameter[7] and observable data from a sample distribution. In addition, he suggests that deductive logic proceeds from the general to the specific to infer this same truth. Furthermore, he suggests that when we have some event that has no deductive arguments available, we may use inductive reasoning to measure its plausibility by working from the particular to the general. He agrees with Richard Threlkeld Cox's (1946) sentiment: that any set of plausibilities that satisfy these desired properties must operate, according to the same rules of probability. Now we can logically revise plausibilities by using the rules of probability, which allows the use of prior truths to project future states of nature.

---

[6]This term is not statistical but practical in nature.

[7]Parameters are statistical reference points such as means and standard deviations, etc.

1.9   Popper's Logic of Scientific Discovery

Karl Popper, the father of deductive scientific reasoning, basically rejects inductive reasoning. For example, he asserts that just because we always see white swans does not mean that there are not non-white ones. He believes only truths can be falsified or rejected, i.e., the rejection of the null hypothesis in classical statistics. Here is where the Bayesian's conflict with the current scientific status quo, as put forth by his idea of falsification and rejection of inductive reasoning. He asserts that we cannot prove but only disprove or falsify the truth. Bayesian probability updating invokes partial truths, which is in direct opposition of his assertion. In current scientific hypothesis testing, we only reject or fail to reject the null; we never prove it as the absolute truth. With Popper's approach, we go away from the truth. With the Bayesian approach, we go toward the truth.

1.10   Frequentist Versus Bayesian (Subjective) Views

Following a Google search for the terms frequentist and Bayesian (subjective), representing the two schools of statistical thought, we quickly note the interest in the latter. This search produced 378,000 results for frequentist and 12,100,000 for Bayesian.[8] The latter etymology began with M. G. Kendall who first used the term to contrast with Bayesian's, whom he called "non-frequentists" (Kendall, 1949). Given the difference between these two schools of thought we will provide some discussion that will differentiate them and provide insights that substantiate them. The frequentist view overshadowed the Bayesian during the first half of the twentieth century. We see the word "Bayesian" first appear in the 1950s and by the 1960s, it became the term preferred by people who sought to escape the limitations and inconsistencies of the frequentist approach to probability theory.

1.10.1   Frequentist to Subjectivist Philosophy

John Maynard Keynes (1921) provides a treatise on the role of the frequentist. His chapter VIII, *The Frequency Theory of Probability*, provides 17 points of insight to the position of subjectivism, which follow:

- Point 1 suggests the difficulty in comparing degrees of probability of the frequentist and offers an alternative theory.

---

[8]We conducted this search on January 1, 2016. When I first published *Strategic Economic Decision-Making: Using Bayesian Belief Networks to Solve Complex Problems* (Grover, 2013), there were only 4,000,000 hits.

- Point 2 suggests a link to frequentist theory back to Aristotle who stated that: "the probability is that which for the most part happens" (p. 92). Keynes traces the frequentist back to Leslie Ellis, whom he suggests, invented the concept that: "If the probability of a given event be correctly determined" … "the event will on a long run of trails tend to recur with frequency proportional to their probability" (p. 93). He also suggests that Venn, in his *"Logic of Chance,"* was an early adopter.
- Point 3 suggests that Venn expresses an interest in probabilities, through an empirically determined series of events, and that one may express probabilities based on experience.
- Point 4 suggests a divergence of probability from frequentist statistics as initiated by Venn.
- Point 5 suggests that Venn's theory is narrowly limited in his exclusion of events that are not certain from the science of probability, which allows us to express statements of frequency. He also suggests that one can derive these probabilities either through inductive, or deductive logic.
- Point 6 suggests two points where we have "induced Venn to regard judgments based on statistical frequency" (p. 97). These are subjectivity, and the inability for us to provide accurate measurements. Venn fails to discuss these in his theory. Therefore, they are not ruled out as being subjective in nature.
- Point 7 suggests that Venn's theory is incomplete because he admits that in most cases we can arrive at statistical frequencies using induction.
- Point 8 suggests that Venn based probabilities on statistical frequencies alone, which were calculable chance. Most importantly, Keynes brings to the discussion the concept of inverse and *a posteriori* probabilities based on statistical grounds.
- Point 9 suggests that Karl Pearson generally agrees with Venn. He suggests a generalized frequency theory that does not regard probability as being identical to statistical frequency.
- Point 10 suggests the use of true proportions as a class of frequencies that measure relative probabilities. Alternatively, "the probability of a proportion always depends upon referring it to some class whose truth-frequency is known within wide or narrow limits" (p. 101). This gives rise to the idea of conditional probability.
- Point 11 suggests criticism of frequency theory based on how one determines the class of reference, which we cannot define as "being the class of proportions of which everything is true is known to be true of the proportion whose probabilities we seek to determine" (p. 103).
- Point 12 suggests a modified view of frequency theory based on the above argument.
- Point 13 suggests one bases the "Additional theorem" (p. 105) on how to derive true proportions that are "independent for knowledge" (p. 106) relative to the given data. This points us to a theorem of "inverse probability" and the use of *a priori* knowledge.
- Point 14 suggests one can base theory of inverse probability on inductive reasoning.
- Points 15–17 suggest additional arguments for inverse theory of probability.

1.10.2   Bayesian Philosophy

While Bayes is the Father of Bayesian inference, we give credit to Pierre-Simon LaPlace for actually deriving the formula, as we see it today. He transitioned probability science from the objective to the subjective school of thought. The former purports that statistical analysis depends only on the assumed model and the analyzed data, and not subjective decisions. Conversely, the subjectivist school did not require objective analysis for hypothesis determination. Fine (2004) reviews Joyce (2008), who suggests that Bayesian probability interprets the concept of probability as a knowledge-base or inductive measure, instead of the frequentist view of an event's probability, as the limit of its relative frequency in a large number of trials. From the Bayesian perspective, the literature presents two views that interpret states of knowledge: the objectivist and the subjectivist schools. The former is an extension of Aristotelian logic, and for the latter, the state of knowledge corresponds to a personal belief. The dominant feature of the subjective school of thought is that one can assign a probability to a hypothesis, which one cannot do as a frequentist. The basis of the theorem, in its simplest form, is its ability to revise the truth, when invoking or conditioning new information. This statement requires a rigorous, alternative approach to probability theory. Its essence is the ability to account for observed information when updating the unobservable. Understanding this concept, is fundamental to learning the theorem. Chapter 2 discusses the evolution of the theorem and BBN in some detail.

1.11   The Identification of the Truth

The above discussion begs for a definition of truth. We define truth as the certainty of an event occurring. From a statistical perspective, even if an event has a 100 % chance of occurring, it could not occur. The theorem allows us to get as close to the truth as possible, when we accept the fact that prior knowledge has intrinsic value. If we reject this form of knowledge; then by default, we remain in the traditionalist camp, and if we accept this, then we enter the Bayesian camp. The latter accepts this knowledge as truth, based on initial assumptions of rational beliefs. The former initially rejects this knowledge as truth through falsification of hypothesis testing.[9] The strength of the Bayesians is that, in searching for the truth, they accept partial truths when updating the *priors*. The strength in the traditionalist is their belief in scientific rigor. Both parties have a dim view of the truth; initially, and each attempts to refine it through deductive and inductive logic.

---

[9]In Bayesian epistemology, $P(B|A)$ is a hypothesis of our initial beliefs. This states that the probability of event B given A is conditioned on a form of evidence, A. (See http://plato. stanford.edu/entries/epistemology-bayesian/ (14 September 2016) for a thorough conversation of this topic.)

Two common Bible verses reference seeing clearly and knowing the truth. "We don't yet see things clearly. We're squinting in a fog, peering through a mist" (1 Corinthians 13:11–13, The Message) and "You will know the truth, and the truth will set you free" (John 8:32, New International Version, Peterson 2016). As Lydia Parks states: "We are on a quest, if we are wise, to find Christ, who is the ultimate truth." If we knew the truth, then there would be no need to search. Thus we search for it, using available information or knowledge, and conditional events, to get closer to the truth. This comes from rigid, systematic research and discoveries, and historical or innate knowledge hidden within traditions and facts. Inductive (Bayesian) logic is allowing the scientific community to overcome traditional constraints, induced by using the deductive logic of falsification.

1.12 Bayes' Theorem: An Introduction

$$P(B_i|A) = \frac{P(A, B_i)}{P(A)}. \tag{1.3}$$

Equation (1.3) is Bayes' theorem in its simplest form. It reads, "the conditional probability of event B_i occurring, given an observable event A is invoked is equal to the joint probability of events A and B_i[10] divided by the marginal probability of event A." Here, B_i is the ith event out of k mutually exclusive and collectively exhaustive events.

We expand this equation, using the chain rule of probability, which states, "the joint probability of events A and B_i is equal to the conditional probability of event A, given the probability of event B_i, times the probability of event B_i." Or equivalently with (1.4):

$$P(A, B_i) = P(A|B_i) \times P(B_i). \tag{1.4}$$

Substituting the chain rule (1.4) into the denominator of Eq. (1.5) yields:

$$P(B_i|A) = \frac{P(A|B_i) \times P(B_i)}{P(A)}. \tag{1.5}$$

Now we can begin to explain the utility of a BBN algorithm. In this example of a two-node BBN, there are two random variables, each with two events: the unobservable event, B_i, and the observable variable, A_i. Let us view each of these events as discrete column vectors consisting of two or more mutually exclusive events. Using a BBN, we specify and illustrate this in Fig. 1.1, where each variable, B_i and A_i, has two mutually exclusive events: B_1, B_2 and A_1, A_2, respectively.

---

[10]This means it is occurring at the same time and on the same trial of the experiment.

Fig. 1.1 Bayesian belief
network

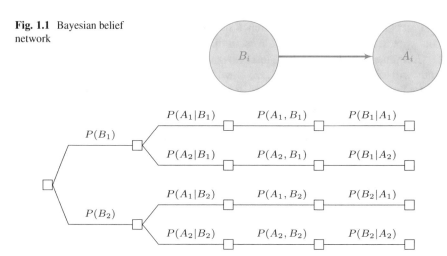

Fig. 1.2 Decision tree for a 2×2 example

To expand on the BBN, Fig. 1.2 represents a decision tree for the two-node
BBN in diagram Fig. 1.1, where we can derive the *prior*, $P(B_i)$, *likelihood* $P(A_i|B_i)$,
joint $P(A_i, B_i)$, and *posterior* probabilities $P(B_i|A_i)$ for the respective nodes. We
decompose this into four subcategories based on whether or not the respective events
of A correctly classify the outcomes of B events.

The decision tree is chained and flows as follows:

- It begins with the *priors*, $P(B_i)$, which are the unconditional (marginal) probabilities that are unobserved and represent our proxy of an initial truth guess,[11]
- then we compute the *likelihoods* $P(A_i|B_i)$,
- the *joints* $P(A_i, B_i)$,
- and finally the *posterior* probabilities $P(B_i|A_i)$.

These *posteriors* are the true probabilities that we seek; they express information
hidden within the *priors* that are not immediately discernable from the data. For
example, the posterior $P(B_1|A_1)$ is the probability of event B_1 occurring, given
that event A_1 had already occurred. Using Bayesian statistics, we can compute any
combination of *posteriors*. Most importantly, we can generalize BBN to represent
n-event models.

Figure 1.3 is a general decision tree process that shows the flow from the *prior*
to *posterior* probabilities.

---

[11]Often we gather these starting points, from observable data. It turns out that this is acceptable,
due to the learning that occurs by the algorithm, when we add multiple observable events to the
BBN.

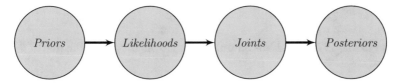

Fig. 1.3 General decision tree process

We will transition now with an example of applying the theorem and BBN in decision-making, using a classical illustration of the Monty Hall, Game Show paradox.

1.13   Classic Illustration-Monty Hall Game Show Paradox

The Monty Hall game show is a classic illustration of how the constructs of the theorem and BBN have ubiquitous utility in decision-making using inductive logic. We base this on the Monty Hall problem (2015), and the American television game show "Let's Make a Deal," named after its original host, Monty Hall. Steve Selvin originally posed the problem in a letter to the American Statistician in 1975. In 1990, it became famous from a reader's letter, quoted in Marilyn vos Savant's "Ask Marilyn" column in Parade magazine.

Our intent is to provide a practical application of the utility of the theorem and BBN using inductive logic and the constructs of discrete mathematics. Below are some characteristics and assumptions of the game show.

1.13.1   *Characteristics*

- an interactive game
- host versus contestant
- there are three doors
- contestant selects one door
- host opens another door showing a goat, but never the car
- a third door remains closed
- host then asks the contestant to switch, or keep the original door

1.13.2   Assumptions

Selvin (1975a, 1975b) and Vos Savant (1991a) explicitly define the role of the host as follows:

- The host must always open a door that the contestant did not select (Mueser and Granberg 1999).
- The host must always open a door to reveal a goat, but never the car.
- The host must always offer the chance to switch between the originally chosen door, and the remaining closed door.
- The host knows which door the car is behind.

Presumably the game show hides the car behind a random door, and if the contestant initially selects the car, then the host's choice of which goat-hiding door to open is random (Krauss and Wang, 2003). Some authors, independently or inclusively, assume the player's initial choice is random as well (Selvin 1975a, 1975b).

1.13.3   Bayesian Belief Network Solution

The question becomes: should the contestant keep the door she or he initially selected, or switch based on the new information provided by the host? To suggest a solution, we have provided a methodology using a fully-specified joint BBN with a 6-Step solution protocol below.

Step 1: Specify the Joint BBN

In the BBN, we have three nodes: (1) Contestant, (2) Car, and (3) Door. There is one contestant, one car, and three doors, and the host is the agent that injects information into the game. We illustrate these three nodes, with paths, in Fig. 1.4.

Step 2: Calculate the Prior Probabilities

In calculating the prior probabilities, the contestant always selects a door at random. Since there are three doors, then each door has an equal chance of being selected. We represent this information in terms of *priors*, as reported in Table 1.1.

Fig. 1.4 BBN nodes &
paths: [*Door → Car|Door →*
Contestant|Contestant → Car]

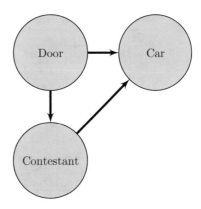

Table 1.1 Monty Hall prior
probabilities

P(Car)

Door 1	Door 2	Door 3	Total
0.333333[a]	0.333333	0.333333	1.000000

[a]0.333333 = $P(Contestant_{Door1})$. This is the
random probability that the car is actually behind
Door 1 with the other doors having the same
probability of selection

Table 1.2 Monty Hall
likelihood probabilities

P(Contestant)

P(Car)	Door 1	Door 2	Door 3	Total
Door 1	0.333333[a]	0.333333	0.333333	1.000000
Door 2	0.333333	0.333333	0.333333	1.000000
Door 3	0.333333	0.333333	0.333333	1.000000

[a]0.333333 = $P(Contestant_{Door1}|Car_{Door1})$. This is the
probability that a contestant randomly selects Door 1 given
the car is actually behind Door 1 (This action is also
random; i.e., the car is actually randomly positioned behind
either Door 1, 2, or 3)

Step 3: Determine the Contestant and Car Likelihood Probabilities

Here are the *conditional* probabilities of the contestant selecting either Door 1, 2,
or 3, given the car is actually behind Door 1, 2, or 3, respectively. We represent this
information in terms of the *likelihoods* reported in Table 1.2.

Step 4: Determine the Host, Contestant, and Car Likelihood Probabilities

We are determining the *conditional* probabilities of the host selecting either Door
1, 2, or 3, given the contestant selects either Door 1, 2, or 3, and that the car is
actually behind one of these doors. Again, we represent this information in terms

Table 1.3 Monty Hall likelihood probabilities

P(Host)

Row	P(Contestant)	P(Car)	Door 1	Door 2	Door 3	Total
1.	Door 1	Door 1	0.000000	0.500000[a]	0.500000[a]	1.000000
2.	Door 1	Door 2	0.000000	0.000000	1.000000	1.000000
3.	Door 1	Door 3	0.000000	1.000000	0.000000	1.000000
4.	Door 2	Door 1	0.000000	0.000000	1.000000	1.000000
5.	Door 2	Door 2	0.500000	0.000000	0.500000	1.000000
6.	Door 2	Door 3	1.000000	0.000000	0.000000	1.000000
7.	Door 3	Door 1	0.000000	1.000000	0.000000	1.000000
8.	Door 3	Door 2	1.000000	0.000000	0.000000	1.000000
9.	Door 3	Door 3	0.500000	0.500000	0.000000	1.000000

[a] Note: 0.500000 = The probability that the host will select either Door 2 or 3 given the contestant selects Door 1 and the car is actually behind Door 1

of *likelihoods*. The selection process is not random, but based on the information provided by the host, and the use of the inductive logic of the contestant.

Below is an explanation of the inductive logic we used to assign *likelihoods* for each row. We report these in Table 1.3.

- Row 1. If the contestant selects Door 1, and the host knows the car is actually behind Door 1, then the only options for the host are to randomly open either Door 2 (50.0 %) or 3 (50.0 %).
- Row 2. If the contestant selects Door 1, and the host knows the car is actually behind Door 2, then the only option for the host is to open Door 3 (100.0 %).
- Row 3. If the contestant selects Door 1, and the host knows the car is actually behind Door 3, then the only option for the host is to open Door 2 (100.0 %).
- Row 4. If the contestant selects Door 2, and the host knows the car is actually behind Door 1, then the only option for the host is to open Door 3 (100.0 %).
- Row 5. If the contestant selects Door 2, and the host knows the car is actually behind Door 2, then the only options for the host are to randomly open either Door 1 (50.0 %) or 3 (50.0 %).
- Row 6. If the contestant selects Door 2, and the host knows the car is actually behind Door 3, then the only option for the host is to open Door 1 (100.0 %).
- Row 7. If the contestant selects Door 3, and the host knows the car is actually behind Door 1, then the only option for the host is to open Door 2 (100.0 %).
- Row 8. If the contestant selects Door 3, and the host knows the car is actually behind Door 2, then the only option for the host is to randomly open Door 1 (100.0 %).
- Row 9. If the contestant selects Door 3, and the host knows the car is actually behind Door 3, then the only options for the host are to open either Door 1 (50.0 %) or 2 (50.0 %).

Table 1.4 Monty Hall joint probabilities

P(Host)					
P(Contestant)	P(Car)	Door 1	Door 2	Door 3	Marginals
Door 1	Door 1	0.000000	0.055556	0.055556[a]	0.111111
Door 1	Door 2	0.000000	0.000000	0.111111	0.111111
Door 1	Door 3	0.000000	0.111111	0.000000	0.111111
Door 2	Door 1	0.000000	0.000000	0.111111	0.111111
Door 2	Door 2	0.055556	0.000000	0.055556	0.111111
Door 2	Door 3	0.111111	0.000000	0.000000	0.111111
Door 3	Door 1	0.000000	0.111111	0.000000	0.111111
Door 3	Door 2	0.111111	0.000000	0.000000	0.111111
Door 3	Door 3	0.055556	0.055556	0.000000	0.111111
Marginals		0.333333	0.333333	0.333333	1.000000

[a]$0.055556 = 0.500000 * 0.333333 * 0.333333 = P(Host_{Door3}, Contestant_{Door1}, Car_{Door1})$. We used the chain rule and *likelihoods* from Tables 1.1, 1.2, and 1.3 using this notation $P(Host_{Door3}|Contestant_{Door1}, Car_{Door1})* P(Contestant_{Door1}|Car_{Door1})* P(Car_{Door1})$ to compute this value

Step 5: Compute the Joint Probabilities

During Step 5, we are determining the probabilities of where this event occurs, at the same time on the same iteration of a random experimental trial, which includes: the host selecting either Door 1, 2, or 3, and the contestant selecting either Door 1, 2, or 3, and the car is actually being behind these doors. We represent this information in terms of the *joints* reported in Table 1.4.

Step 6: Compute the Posterior Probabilities

Now we can answer our initial question: Is it better for the contestant to keep her or his initial selected door or switch? We invoke the constructs of the theorem using inductive logic, and calculate the *posteriors*. We determine the *conditional* probabilities of the car actually being behind either Door 1, 2, or 3, given that the contestant selects either Door 1, 2, or 3, and the host selects the door that the car is not behind. We do this by drawing from the *joints* we computed in Table 1.4. The logic follows for each row in Table 1.5.

- Row 1. The *conditional* probabilities do not exist. The host never opens a door that the car is behind, or that the contestant selects.
- Row 2. The probability of the car being behind Door 1, given the contestant selects Door 1, and the host opens Door 2 is 33.3 % and 66.7 % for Door 3.

We add the probabilities of Door 2 to Door 3, because the host reveals, that the car is not behind that door.[12]

- Row 3. The probability of the car being behind Door 1, given the contestant selects Door 1, and the host opens Door 3 is 33.3 % and 66.7 % for Door 2. We add the probabilities of Door 3 to Door 2 because the host reveals, that the car is not behind that door.[13]

- Row 4. The probability of the car being behind Door 2, given the contestant selects Door 2, and the host opens Door 1 is 33.3 % and 66.7 % for Door 3. We add the probabilities of Door 1 to Door 3, because the host reveals that the car is not behind that door.[14]

- Row 5. The *conditional* probabilities do not exist. The host never opens a door that the car is behind, or that the contestant selects.

- Row 6. The probability of the car being behind Door 1, given the contestant selects Door 2, and the host opens Door 3 is 66.7 % and 33.3 % for Door 2. We add the probabilities of Door 3 to Door 1, because the host reveals that the car is not behind that door.[15]

- Row 7. The probability of the car being behind Door 2, given the contestant selects Door 3, and the host opens Door 1 is 66.7 % and 33.3 % for Door 3. We add the probabilities of Door 1 to Door 3, because the host reveals that the car is not behind that door.[16]

- Row 8. The probability of the car being behind Door 1, given the contestant selects Door 3, and the host opens Door 2 is 66.7 % and 33.3 % for Door 3. We add the probabilities of Door 2 to Door 1, because the host reveals that the car is not behind that door.[17]

- Row 9. The *conditional* probabilities do not exist. The host never opens a door the car is behind, or that the contestant selects.

---

[12]For Door 1, we compute this as: $P(Car_{Door1}|Contestant_{Door1}, Host_{Door2}) = P(Car_{Door1}, Contestant_{Door1}, Host_{Door2})/P(Contestant_{Door1}, Host_{Door2}) = 0.333333 = 0.055556/(0.055556 + 0.000000 + 0.111111)$.

[13]For Door 1, we compute this as: $P(Car_{Door1}|Contestant_{Door1}, Host_{Door3}) = P(Car_{Door1}, Contestant_{Door1}, Host_{Door3})/P(Contestant_{Door1}, Host_{Door3}) = 0.333333 = 0.055556/(0.055556 + 0.111111 + 0.000000)$.

[14]For Door 2, we compute this as: $P(Car_{Door2}|Contestant_{Door2}, Host_{Door1}) = P(Car_{Door2}, Contestant_{Door2}, Host_{Door1})/P(Contestant_{Door2}, Host_{Door1}) = 0.333333 = 0.055556/(0.000000 + 0.055556 + 0.111111)$.

[15]For Door 1, we compute this as: $P(Car_{Door1}|Contestant_{Door2}, Host_{Door3}) = P(Car_{Door1}, Contestant_{Door2}, Host_{Door3})/P(Contestant_{Door2}, Host_{Door3}) = 0.666667 = 0.111111/(0.111111 + 0.055556 + 0.000000)$.

[16]For Door 2, we compute this as: $P(Car_{Door2}|Contestant_{Door3}, Host_{Door1}) = P(Car_{Door2}, Contestant_{Door3}, Host_{Door1})/P(Contestant_{Door3}, Host_{Door1}) = 0.666667 = 0.111111/(0.000000 + 0.111111 + 0.055556)$.

[17]For Door 1, we compute this as: $P(Car_{Door1}|Contestant_{Door3}, Host_{Door2}) = P(Car_{Door1}, Contestant_{Door3}, Host_{Door2})/P(Contestant_{Door3}, Host_{Door2}) = 0.666667 = 0.111111/(0.111111 + 0.000000 + 0.055556)$.

We report the posterior probabilities in Table 1.5:

Table 1.5 Monty Hall posterior probabilities

P(Car)						
Row	P(Contestant)	P(Host)	Door 1	Door 2	Door 3	Total
1.	Door 1	Door 1	NA	NA	NA	NA
2.	Door 1	Door 2	0.333333	0.000000	0.666667[a]	1.000000
3.	Door 1	Door 3	0.333333	0.666667	0.000000	1.000000
4.	Door 2	Door 1	0.000000	0.333333	0.666667	1.000000
5.	Door 2	Door 2	NA	NA	NA	NA
6.	Door 2	Door 3	0.666667	0.333333	0.000000	1.000000
7.	Door 3	Door 1	0.000000	0.666667	0.333333	1.000000
8.	Door 3	Door 2	0.666667	0.000000	0.333333	1.000000
9.	Door 3	Door 3	NA	NA	NA	NA

[a] $0.666667 = 0.111111/(0.055556 + 0.000000 + 0.111111) = P(Car_{Door3}|Contestant_{Door1}, Host_{Door2})$. This is the probability that the can is behind Door 3 given the contestant selects Door 1 and the Host opens Door 2

1.13.4 Conclusions

As illustrated in the *posterior* probabilities in Table 1.5, it is always better to switch when you are presented with new information than to stay with your initial selection, when new information suggests or weights the evidence against your initial assumptions. There is a 2/3 chance, or 66.7 %, that if the contestant switches, then the car will be behind the new door.[18]

References

Bolstad, W. M. (2007). *Introduction to Bayesian statistics* (2nd ed.). Hoboken, NJ: Wiley.

Cox, R. T. (1946). Probability, frequency, and reasonable expectation. *American Journal of Physics, 14*, 1–13.

Fine, T. L. (2004). The "only acceptable approach" to probabilistic reasoning. In E. T. Jaynes (Ed.), *Probability theory: The logic of science*. Cambridge: Cambridge University Press; (2003). *SIAM News, 37*(2), 758.

Grover, J. (2013) *Strategic economic decision-making: Using Bayesian belief networks to solve complex problems. SpringerBriefs in Statistics* (Vol. 9). New York: Springer Science+Business Media. doi:10.1007/978-1-4614-6040-4_1.

---

[18]Ruling out rows 1, 5, and 9 are key to solving this problem.

Jaynes, E. T. (1995). *Probability theory: The logic of science.* http://shawnslayton.com/open/Probability%2520book/book.pdf. Accessed 26 June 2012.

Joyce, J. (2008). In E. N. Zalta (Ed.),"Bayes' theorem". *The Stanford encyclopedia of philosophy* (Fall 2008 ed.). http://plato.stanford.edu/archives/fall2008/entries/bayes-theorem/. Accessed 1 July 2012.

Kendall, M. G. (1949). On the reconciliation of theories of probability. *Biometrika (Biometrika Trust), 36*(1/2), 101–116. doi:10.1093/biomet/36.1-2.101.

Keynes, J. M. (1921). *A treatise on probability*: Macmillan and Co. Universal digital library. Collection: Universallibrary. Retrieved from http://archive.org/details/treatiseonprobab007528mbp. Accessed 15 June 2012.

Krauss, S., & Wang, X. T. (2003). The psychology of the Monty Hall problem: Discovering psychological mechanisms for solving a tenacious brain teaser. *Journal of Experimental Psychology: General, 132*(1), 3–22. doi:10.1037/0096-3445.132.1.3. Retrieved 14 September 2016 from http://usd-apps.usd.edu/xtwanglab/Papers/MontyHallPaper.pdf

Monty Hall problem (2015, August 28). Retrieved September 4, 2015 from http://en.wikipedia.org/wiki/Monty_Hall_problem

Mueser, P. R., & Granberg, D. (1999). The Monty Hall Dilemma revisited: Understanding the interaction of problem definition and decision making. Working Paper 99–06, University of Missouri. Retrieved 6 November 2016 from http://econpapers.repec.org/paper/wpawuwpex/9906001.htm.

New International Version. Biblica, 2011. Bible Gateway. Web 14 Sep. 2016.

Peterson, E. H. *The Message. Bible Gateway*. Web. 14 Sep. 2016.

Selvin, S. (1975a, February). A problem in probability (letter to the editor). *American Statistician, 29* (1), 67. JSTOR 2683689.

Selvin, S. (1975b, August). On the Monty Hall problem (letter to the editor). *American Statistician, 29*(3), 134. JSTOR 2683443.

vos Savant, M. (1990a, September 9). Ask Marilyn. Parade Magazine: 16.

vos Savant, M. (1991a, February 17). Ask Marilyn. Parade Magazine: 12.

Chapter 2
Literature Review

2.1 Introduction to the Bayes' Theorem Evolution

This section provides a literature review of Bayes' theorem and Bayesian Belief Networks (BBN). The concept of the theorem[1] begins with a series of publications beginning with the *"Doctrine of Chances"* by Abraham de Moivre during the period of 1718–1756 (Schneider 2005). Historians have named the theorem after the Reverend Thomas Bayes[2] (1702–1761), who studied how to compute a distribution for the parameter of a binomial distribution. His friend, Richard Price, edited and presented the work in 1763, after his death, as *"An Essay towards solving a Problem in the Doctrine of Chances"* (Bayes and Price 1763). Of particular importance is his Proposition 9. Of greater importance is Bayes' original idea of using a "Starting Guess" for a parameter of interest. This ignites the science of inverse probability and the beginning of a new school of probability thought. We see the different schools linked to philosophical approaches such as "Classical" statistics from R.A. Fisher's *p*-values, Aris Spanos Jerzy Neyman's deductive hypothesis tests, and the Popperian view of science, which suggests that a hypothesis is made, and then it is tested and can only be rejected or falsified, but never accepted (Lehmann 1995). The Bayesian epistemology[3] runs contrary to these schools of thought.

In 1774, Pierre-Simon LaPlace publishes his first version of inverse probability following his study of de Moivre's *"Doctrine of Chance,"* presumably the 1756 version. His final rule was in the form we still use today in Eq. (2.1):

---

[1] We obtained most of the facts on the evolution of the theorem herein from McGrayne (2011).

[2] There is still debate on the true author of Bayes' theorem. Some give the honor to Pierre-Simon LaPlace, following his 1774 publication of a similar theorem.

[3] See Joyce's comments on Bayesian epistemology, for a complete discussion (Joyce 2008).

© Springer International Publishing AG 2016
J. Grover, *The Manual of Strategic Economic Decision Making*,
DOI 10.1007/978-3-319-48414-3_2

$$P(C|E) = \frac{P(E|C)P_{prior}(C)}{\sum P(E|C')P_{prior}(C')},$$ (2.1)

where $P(C|E)$ is the probability of a hypothesis C (Cause), given data or information E (Evidence), which is equal to the probability of new information, $P(E|C)$, times the prior information $P_{prior}(C)$, divided by the sum of the probabilities of the data of all possible hypotheses $\sum P(E|C')P_{prior}(C')$. In the late 1870s early 1880s, Charles Sanders Peirce championed frequency-based probability, which launches this stream of empirical thought. In 1881, George Chrystal challenges Laplace's idea of the theorem, and declares that the laws of inverse probability are dead. Towards the end of the nineteenth century we begin to see some utility of the theorem when the French mathematician and physicist Henri Poincare' invokes it during a military trial following the Dreyfus affair of 1899. He suggests Dreyfus, a Jewish, French army officer, was not a German spy.

2.1.1 Early 1900s

In 1918, Edward C. Molina, a New York City engineer and self-taught mathematician, uses the theorem to evaluate the economic value of automating the Bell telephone system with call data, thus preventing them from bankruptcy. Albert Wurts Whitney, a Berkley insurance mathematics expert, uses the theorem to establish a form of social insurance with optimized premiums.

2.1.2 1920s–1930s

In 1926, Sir Harold Jeffreys, the Father of modern Bayesian statistics, uses the theorem to infer that the Earth's core is liquid, and Frank P. Ramsey, English mathematician, and philosopher, suggests how one can make quantified uncertain decisions using personal beliefs when making a wager. In 1933, Andrey Kolmogorov, a Soviet mathematician, suggests the use of the theorem, as a method of firing back at a German artillery bombardment of Moscow, using Bertrand's Bayesian firing system. In 1936, Lowell J. Reed, a medical researcher at Johns Hopkins University, uses the theorem to determine the minimum amount of radiation required to cure cancer patients, while causing the least amount of damage. In 1938, Erik Essen-Möller, a Swedish professor of genetics and psychiatry, develops an index of probability for paternity testing, which was used until the advent of DNA testing. In 1939, Harold Jeffreys, a geologist, publishes his theory of probability that uses the theorem, as the only method to conduct scientific experiments with subjective probabilities.

2.1.3   1940s–1950s

In 1941, Alan Mathison Turning, the father of the modern computer, invents a Bayesian system of Bankurismus using banded Banburg strips that uses sequential analysis to break the German Enigma Coding Machine. In 1942, Kolmogorov introduces firing dispersion theory, which is a Bayesian scoring system using a 50–50 guess for aiming artillery, and again; Turing invents Bayesian Turingismus to deduce the patterns of cams surrounding the Tunny-Lorenz machine. He uses "gut feels" as *prior* probabilities. From 1943 to 1944, Max Newman, a British mathematician, and code-breaker, invents the Colossus I and II machines and intercepts a message that Hitler gave Rommel, ordering a delay of his attack in Normandy. In 1945, Ian Cassels, Jimmy Whitworth, and Edward Simpson, all cryptanalysts, use the theorem to break Japanese code during WWII, and John Gleason, a cryptanalyst, uses the theorem to break Russian code, during the Cold War era. In 1947, Arthur L. Baily, an insurance actuary, resurrects Bayes' theorem, and demands the legitimate use of *prior* probabilities, using Bible verses as references. In 1950, he reads his work on Credibility procedures, during an actuarial society banquet, citing LaPlace's form of the theorem, and the combination of prior knowledge, with observed data. In 1951, Jerome Cornfield, a history major, working at the National Institute of Health, uses the theorem to provide a solid theoretical link that smoking does cause cancer. In 1954, Jimmie Savage, a University of Chicago statistician, publishes his revolutionary book, the *"Foundations of Statistics,"* which extends Frank Ramsey's use of the theorem for making inferences and decision-making. In 1955, L.H. Longly-Cook, a chief actuary, predicts the first U.S. catastrophic aviation disaster, which involve two planes colliding in mid-air. This allows insurance companies to raise rates prior to this event. Hans Bühlmann, a mathematics professor, extends Baily's Bayes' philosophy, and publishes a general Bayesian theory of credibility. In 1958, Albert Madansky, a statistician, writes a summary of the theorem in the RANDS Corps: *"On the Risk of an Accidental or Unauthorized Nuclear Detonation,"* suggesting a probability greater than zero that this event could occur. Finally, in 1959, Robert Osher Schlaifer, a Harvard University's statistician, publishes *"Probability and Statistics for Business Decisions, An Introduction to Managerial Economics under Uncertainty."*

2.1.4   1960s–Mid 1980s

In 1960, Morris H. DeGroot, a practitioner, publishes the first international text on Bayesian decision theory. Frederick Mosteller, Harvard University professor, and David L. Wallace, University of Chicago statistician, evaluate the 12 unknown authors of the Federalist papers, using the theorem to identify Madison as the correct author. John W. Tukey, a Princeton statistic's professor, predicts Nixon as the winner of the Nixon-Kennedy presidential elections for NBC, using their mainframe computers and Bayesian-like code. In 1961, Homer Warner, a pediatric heart surgeon, develops the first computerized program for diagnosis of diseases using the theorem. Robert Osher Schlaiter and Howard Raiffa, two Harvard University business professors, publish *"Applied Statistical Decision Theory,"* a classical work using the theorem, and this charters the future direction for Bayesian theory. In 1968, John Piña Craven, civilian chief scientist, and Frank A. Andrews, Navy Captain (R), locate the sunken submarine, the *U.S.S. Scorpion* using Bayesian search techniques.

In 1974, Norman Carl Rasmussen, a physicist, and engineer uses Raiffas' decision trees (Raiffa 2012) to weigh the risks of meltdowns in the nuclear-power industry for the U.S. Nuclear Regulatory Commission (NRC) (Fienberg 2008). The NRC halts the study, due to his inclusion of the theorem is resurrected it following the 1979 Three Mile Island incident. In 1975, Lawrence D. Stone, a Daniel H. Wagner Associates employee, publishes a *"Theory of Optimal Search"* using Bayesian techniques, following his participation in locating the *U.S.S. Scorpion,* and the NRC gives him the invitation to publish his findings. In 1976, Harry C. Andrews, a digital image processor, publishes *"Digital Image Restoration."* This uses Bayes' inference to restore nuclear weapons testing images from activity at Los Alamos National Laboratories. Finally, in 1983, Teledyne Energy Systems uses hierarchical methods to estimate shuttle failure at 35:1, when NASA estimated it as 100,000:1.[4]

2.1.5   Mid 1980s to Today

The mid-1980s mark the beginning of the BBN evolution. In 1985, Judea Pearl, computer scientist, publishes the seminal work on BBN, *"Bayesian Networks: A Model of Self Activated Memory for Evidential Reasoning"* (Pearl, 1985) to guide the direction of BBN, using discrete random variables and distributions. The following empirical studies are representative of peer review extensions to his work from 2005, to the present.[5,6]

2.1.5.1   Financial Economics, Accounting, and Operational Risks

BBN studies in these areas include: gathering information in organizations (Calvo-Armengol and Beltran 2009), conducting Bayesian learning in social networks

---

[4]In 1986, the Challenger explodes.

[5]We queried these through the Social Science Citation Index Web of Science ® (Reuters 2012).

[6]Unless otherwise reference, certain data included herein are derived from the Web of Science ® prepared by Thomson Reuters ®, Inc. (Thomson ®), Philadelphia, Pennsylvania, USA: # Copyright Thomson Reuters ® 2012. All rights reserved.

(Acemoglu et al. 2011), processing information (Zellner 2002), evaluating games and economic behavior (Mannor and Shinikin 2008), economic theory and market collapse (Gunay 2008), determining accounting errors (Christensen 2010), evaluating operational risk in financial institutions (Neil et al. 2009), determining the valuation of contingent claims with mortality, and evaluating interest rate risks using mathematics and computer modeling techniques (Jalen and Mamon 2009).

2.1.5.2 Safety, Accident Analysis, and Prevention

BBN studies in these areas include: epidemiology, environment, human safety, injury and accidents, road design, and urban settings (DiMaggio and Li 2012), evaluating infant mortality, deprivation, and proximity to polluting industrial facilities (Padilla et al. 2011), human-centered safety analysis of prospective road designs (Gregoriades et al. 2010), predicting real-time crashes on the basic freeway segments of urban expressways (Hossain and Muromachi 2012) and crash counts by severity (Ma et al. 2008), evaluating the effects of osteoporosis on injury risk in motor-vehicle crashes (Rupp et al. 2010), and workplace accidents caused by falls from a height (Martin et al. 2009).

2.1.5.3 Engineering and Safety

BBN studies in these areas include: incorporating organizational factors into probabilistic risk assessment of complex socio-technical systems (Mohaghegh et al. 2009), predicting workloads for improved design and reliability of complex systems (Gregoriades and Sutcliffe 2008b), evaluating a methodology for assessing transportation network terrorism risk, with attacker and defender interactions (Murray-Tuite and Fei 2010), evaluating individual safety and health outcomes, in the construction industry (McCabe et al. 2008), evaluating risk and assessment methodologies at work sites (Marhavilas et al. 2011), quantifying schedule risk in construction projects (Luu et al. 2009), and studying emerging technologies, that evaluate railroad transportation of dangerous goods (Verma 2011).

2.1.5.4 Risk Analysis

BBN studies in this area include: developing a practical framework for the construction of a biotracing model, as it applies to salmonella in the pork slaughterchain (Smid et al. 2011), assessing and managing risks posed by emerging diseases (Walshe and Burgman 2010), identifying alternative methods for computing the sensitivity of complex surveillance systems (Hood et al. 2009), assessing uncertainty in fundamental assumptions and associated models for cancer risk assessment (Small 2008), modeling uncertainty using model performance data (Droguett and

Mosleh 2008), using Bayesian temporal source attribution, to evaluate foodborne zoonoses (Ranta et al. 2011), and developing posterior probability models, in risk-based integrity modeling (Thodi et al. 2010).

2.1.5.5 Ecology

BBN studies in this area include: studying marine ecology to evaluate integrated modeling tools for marine spatial management (Stelzenmuller et al. 2011), integrating fuzzy, cognitive mapping, in a livelihood vulnerability analysis (Murungweni et al. 2011), water resources management in Spain (Zorrilla et al. 2010), irrigation management in the Indian Himalayas (Saravanan 2010), evaluating feral cat management options (Loyd and DeVore 2010), conducting an integrated analysis of human impact on forest biodiversity in Latin America (Newton et al. 2009), and integrating biological, economic, and sociological knowledge to evaluate management plans for Baltic salmon (Levontin et al. 2011).

2.1.5.6 Human Behavior

BBN studies in this area include: evaluating psychological and psychiatric factors in decision-making on ambiguous stimuli, such as prosody, by subjects suffering from paranoid schizophrenia, alcohol dependence (Fabianczyk 2011), studying substance use to estimate population prevalence from the Alcohol Use Disorders Identification Test scores (Foxcroft et al. 2009), evaluating the role of time and place, in the modeling of substance abuse patterns following a mass trauma (Dimaggio et al. 2009), and affective disorders, on applied non-adult dental age assessment methods in identifying skeletal remains (Heuze and Braga 2008).

2.1.5.7 Behavioral Sciences and Marketing

BBN studies in these areas include: (1) Behavioral Sciences: analyzing adaptive management and participatory systems (Smith et al. 2007), evaluating human behavior in the development of an interactive, computer-based interface, to support the discovery of individuals' mental representations and preferences in decisions problems as they relate to traveling behavior (Kusumastuti et al. 2011), determining semantic coherence (Fisher and Wolfe 2011), conducting a behavioral and brain science study, to evaluate base-rates in ordinary people (Laming 2007), evaluating the implications of natural sampling, in base-rate tasks (Kleiter 2007), evaluating a probabilistic approach to human reasoning, as a précis of Bayesian rationality (Oaksford and Chater 2009), and conducting an environmental and behavioral study to model and measure individuals' mental representations of complex spatio-temporal, decision problems (Arentze et al. 2008). (2) Marketing: evaluating marketplace behavior (Allenby 2012), modeling a decision-making aid for intelligence and marketing analysts (Michaeli and Simon 2008), and investigating endogeneity bias in marketing (Liu et al. 2007).

2.1.5.8 Decision Support Systems (DSS) with Expert Systems (ES) and Applications, Information Sciences, Intelligent Data Analysis, Neuroimaging, Environmental Modeling and Software, and Industrial Ergonomics

BBN studies in these areas include: (1) DDS with ES and Applications: aiding the diagnosis of dementia (Mazzocco and Hussain 2012), determining customer churn analysis in the telecom industry of Turkey (Kisioglu and Topcu 2011), conducting a customer's perception risk analysis in new-product development (Tang et al. 2011), assessing critical success factors for military decision support (Louvieris et al. 2010), predicting tourism loyalty (Hsu et al. 2009), Korean, box-office performance (Lee and Chang 2009), using data mining techniques to detect fraudulent financial statements (Kirkos et al. 2007 and Ngai et al. 2011). (2) Information Sciences: evaluating intelligent and adaptive car interfaces (Nasoz et al., 2010). (3) Intelligent Data Analysis: evaluating automatic term recognition (Wong et al., 2009), and a socio-technical approach to business process simulation (Gregoriades and Sutcliffe, 2008a). (4) Neuroimaging: conducting multi-subject analyses with dynamic causal modeling (Kasess et al., 2010). (5) Environmental Modeling and Software: evaluating perceived effectiveness of environmental DDS in participatory planning, using small groups of end-users (Inman et al., 2011), modeling linked economic valuation and catchment (Kragt et al., 2011). (6) Industrial Ergonomics: exploring diagnostic medicine using DDS (Lindgaard et al., 2009).

2.1.5.9 Cognitive Science

BBN studies in this area include: evaluating the role of coherence in multiple testimonies (Harris and Hahn, 2009) and a learning diphone-based segmentation (Daland and Pierrehumbert, 2011), evaluating the efficiency in learning and problem solving (Hoffman and Schraw, 2010), evaluating the base-rate scores of the Millon Clinical Multiaxial Inventory-III (Grove and Vrieze, 2009), evaluating spatial proximity and the risk of psychopathology following a terrorist attack (DiMaggio et al., 2010), evaluating actuarial estimates of sexual recidivism risk (Donaldson and Wollert, 2008), and evaluating sexually violent predator evaluations (Wollert, 2007).

2.1.5.10 Medical, Health, Dental, and Nursing

BBN studies in these areas include: (1) Medical: evaluating the risk of tuberculosis infection for individuals lost to follow-up evaluations (Martinez et al., 2008), and assessing differences between physicians' realized anticipated gains from electronic health record adoption (Peterson et al., 2011). (2) Health: evaluating socioeconomic inequalities in mortality in Barcelona (Cano-Serral et al., 2009), estimating race and associated disparities, where administrative records lack self-reporting of race (Elliott et al., 2008), and facilitating uncertainty in economic evaluations of patient level data (McCarron et al., 2009). (3) Dental: combining surveillance and expert evidence of viral hemorrhagic septicemia freedom (Gustafson et al., 2010), and investigating dentists' and dental students' estimates of diagnostic probabilities (Chambers et al., 2010). (4) Nursing: evaluating affective disorders in postnatal depression screening (Milgrom et al., 2011), estimating coronary heart disease risk in asymptomatic adults (Boo et al., 2012), determining the efficacy of T'ai Chi (Carpenter et al., 2008), and evaluating diagnostic test efficacy (Replogle et al., 2009).

2.1.5.11   Environmental Studies

BBN studies in this area include: identifying potential compatibilities and conflicts between development and landscape conservation (McCloskey et al., 2011), evaluating longer-term mobility decisions (Oakil et al., 2011), assessing uncertainty, in urban simulations (Sevcikova et al., 2007), modeling land-use decisions, under conditions of uncertainty (Ma et al., 2007), determining the impact of demographic trends on future development patterns, and the loss of open space in the California Mojave Desert (Gomben et al., 2012), determining a methodology to facilitate compliance with water quality regulations (Joseph et al., 2010), and using participatory object-oriented Bayesian networks and agro-economic models for groundwater management in Spain (Carmona et al., 2011).

2.1.5.12   Miscellaneous: Politics, Geriatrics, Space Policy, and Language and Speech

BBN studies in these areas include: (1) Politics: an evaluation of partisan bias and the Bayesian ideal, in the study of public opinion (Bullock, 2009). (2) Geriatrics: an evaluation of the accuracy of spirometry, in diagnosing pulmonary restriction in elderly people (Scarlata et al., 2009). (3) Space Policy: the value of information in methodological frontiers and new applications for realizing asocial benefit (Macauley and Laxminarayan, 2010). (4) Language and Speech: quantified evidence, in forensic authorship analysis (Grant, 2007), and causal explanation and fact mutability in counterfactual reasoning (Dehghani et al., 2012).

2.1.5.13 Current Government and Commercial Users of Bayesian Belief Networks

The following list reports the utility of the theorem in current business, government, and in commerce:

- Analyzing roadway safety measures (Schultz et al., 2011)
- Applications in land operations (Starr and Shi, 2004)
- Building process improvement business cases (Linders, 2009)
- Comparing public housing and housing voucher tenants (The U.S. Department of Housing and Urban Development) (Mast, 2012)
- Conducting social network analysis (Koelle et al. n.d.)
- Conducting unified, flexible and adaptable analysis of misuses and anomalies, in network intrusion detection and prevention systems (Bringas, 2007)
- Designing food (Corney, 2000)
- Evaluating the risk of erosion in peat soils (Aalders et al., 2011)
- Evaluating U.S. county poverty rates (The U.S. Census Bureau) (Asher and Fisher, 2000)
- Executing cognitive social simulation from a document corpus (The Modeling, Virtual Environments, and Simulation Institute) (McKaughan et al., 2011)
- Identifying military clustering problem sets (BAE Systems) (Sebastiani et al., 1999)
- Identifying potential compatibilities and conflicts, between development and landscape conservation (McCloskey et al., 2011)
- Improving Attrition Rates in the M1A1/M1A2 Master Gunner Course (U.S. Army) (Zimmerman et al., 2010)
- Investigating engineering design problems (The U.S. Department of Energy) (Swiler, 2006)
- Investigating the relationships between environmental stressors and stream condition (Allan et al., 2012)
- Measuring the internal dosimetry of uranium isotopes (The Los Alamos National Laboratory) (Little et al., 2003)
- Measuring neighborhood quality, with survey data (The U.S. Department of Housing and Urban Development) (Mast, 2010)
- Modeling the reliability of search and rescue operations within The United Kingdom Coastguard (maritime rescue) coordination centres (Norrington et al., 2008)
- Optimizing and parameter estimation in environmental management (Vans, 1998)
- Predicting long-term shoreline change due to sea-level rise (The U.S. Geological Survey Data Series) (Gutierrez et al., 2011)
- Predicting the impacts of commercializing non-timber forest products on livelihoods, ecology, and society (Newton et al., 2006)
- Predicting the reliability of military vehicles (Neil et al., 2001)

- Ranking of datasets (U.S. Government)[7]
- Use in U.S. Government public policy and government settings including: city growth in the areas of census-taking and small area estimation, U.S. election night forecasting, U.S. Food and Drug Administration studies, assessing global climate change, and measuring potential declines in disability among the elderly (Fienberg, 2011)
- Analyzing information system network risk (Staker, 1999)

2.1.6   Trademarked Uses of Bayesian Belief Networks

A quick search of "Bayes" and "theorem" at the U.S. Patent website[8] provides the latest 45 patents in 2016, that use Bayes' theorem in their inventions. It is clear that the theorem has great utility here as well. The list below reports these with the Patent No. and title for 2016. They include areas such as signal transmissions, cloud computing, pulse, pattern recognition, imagery, classifications, social media, energy, neurology, malware, data visualization, and speech recognition.

- Analyzing device similarity
- Bathymetric techniques, using satellite imagery
- Classifying computer files as malware or whiteware
- Content recommendation for groups
- Decomposition apparatus and method for refining composition of mixed pixels in remote sensing images
- Depth-aware, blur kernel estimation method for iris deblurring
- Detecting and modifying facial features of persons in images
- Detecting network anomalies by probabilistic modeling of argument strings with Markov Chains
- Device and method, for identifying anomalies on instruments
- Dynamic clustering of nametags, in an automated, speech recognition system
- Filtering confidential information, in voice and image data
- Filtering road traffic data, from multiple data sources
- Game theoretic prioritization system and method
- Interactive display of computer aided detection results, in combination with quantitative prompts
- Just-in-time analytics on large file systems and hidden databases
- Knowledge discovery from citation networks
- Method and system for estimating information related to a vehicle pitch and/or roll angle
- Method and system of identifying users, based upon free text keystroke patterns

---

[7]See Data.gov. Data.gov FAQ (2012).

[8]http://patft.uspto.gov/netahtml/PTO/index.html.

- Method for automatic, unsupervised classification of high-frequency oscillations, in physiological recordings
- Method for coding pulse vectors using statistical properties
- Method of identifying materials from multi-energy X-rays
- Methods and systems architecture, to virtualize energy functions and processes into a cloud based model
- Methods and systems, for determining assessment characters
- Methods for non-invasive prenatal, ploidy calling
- Methods for providing operator support, utilizing a vehicle telematics service system
- Multi-contrast image reconstruction, with joint Bayesian compressed sensing
- Neurological prosthesis
- Pattern recognizing engine
- Peer-to-peer data storage
- Posterior probability of diagnosis index
- Power consumption management
- Pre-coding method and pre-coding device
- Precoding method and transmitting device
- Probabilistic methods and apparatus to determine the state of a media device
- Probabilistic registration of interactions, actions or activities from multiple views
- Ranking authors in social media systems
- Relationship information expansion apparatus, relationship information expansion method and program
- Signal generating method and signal generating device
- System and method for probabilistic name matching
- System and method for providing information tagging in a networked system
- Systems and methods for routing a facsimile confirmation based on content
- Transmission method, transmission device, reception method, and reception device
- Transmission method, transmitter apparatus, reception method and receiver apparatus
- Visualization of paths using GPS data
- Wide band, clear air scatter, Doppler radar

References

Aalders, I., Hough, R. L., & Towers, W. (2011). Risk of erosion in peat soils – an investigation using Bayesian belief networks. *Soil Use and Management, 27*(4), 538–549. doi:10.1111/j.1475-2743.2011.00359.x.

Acemoglu, D., Dahleh, M. A., Lobel, I., & Ozdaglar, A. (2011). Bayesian learning in social networks. *Review of Economic Studies, 78*(4), 1201–1236. doi:10.1093/restud/rdr004.

Allan, J. D., Yuan, L. L., Black, P., Stockton, T., Davies, P. E., Magierowski, R. H., et al. (2012). Investigating the relationships between environmental stressors and stream condition using Bayesian belief networks. *Freshwater Biology, 57*, 58–73. doi:10.1111/j.1365-2427.2011.02683.x.

Allenby, G. M. (2012). Modeling marketplace behavior. *Journal of the Academy of Marketing Science, 40*(1), 155–166. doi:10.1007/s11747-011-0280-3.

Arentze, T. A., Dellaert, B. G. C., & Timmermans, H. J. P. (2008). Modeling and measuring individuals' mental representations of complex spatio-temporal decision problems. *Environment and Behavior, 40*(6), 843–869. doi:10.1177/0013916507309994.

Asher, J., & Fisher, R. (2000). Alternative scaling parameter functions in a hierarchical Bayes model of U.S. county poverty rates. In *Proceedings of the Survey Research Methods Section, ASA*.

Bayes, T., & Price, R. (1763). An essay towards solving a problem in the doctrine of chance. By the late Rev. Mr. Bayes, communicated by Mr. Price, in a letter to John Canton, M. A. and F. R. S. *Philosophical Transactions of the Royal Society of London, 53*, 370–418. doi:10.1098/rstl.1763.0053.

Boo, S., Waters, C. M., & Froelicher, E. S. (2012). Coronary heart disease risk estimation in asymptomatic adults. *Nursing Research, 61*(1), 66–69. doi:10.1097/NNR.0b013e31823b1429.

Bringas, P. G. (2007). Intensive use of Bayesian belief networks for the unified, flexible and adaptable analysis of misuses and anomalies in network intrusion detection and prevention systems. *Database and Expert Systems Applications, 2007. DEXA '07. 18th International Workshop on, 3–7 September 2007* (pp. 365–371). doi:10.1109/dexa.2007.38.

Bullock, J. G. (2009). Partisan bias and the Bayesian ideal in the study of public opinion. *Journal of Politics, 71*(3), 1109–1124. doi:10.1017/s0022381609090914.

Calvo-Armengol, A., & Beltran, J. D. (2009). Information gathering in organizations: Equilibrium, welfare, and optimal network structure. *Journal of the European Economic Association, 7*(1), 116–161.

Cano-Serral, G., Azlor, E., Rodriguez-Sanz, M., Pasarin, M. I., Martinez, J. M., Puigpinos, R., et al. (2009). Socioeconomic inequalities in mortality in Barcelona: A study based on census tracts (MEDEA project). *Health & Place, 15*(1), 186–192. doi:10.1016/j.healthplace.2008.04.004.

Carmona, G., Varela-Ortega, C., & Bromley, J. (2011). The use of participatory object-oriented Bayesian networks and agro-economic models for groundwater management in Spain. *Water Resources Management, 25*(5), 1509–1524. doi:10.1007/s11269-010-9757-y.

Carpenter, J., Gajewski, B., Teel, C., & Aaronson, L. S. (2008). Bayesian data analysis: Estimating the efficacy of t'ai chi as a case study. *Nursing Research, 57*(3), 214–219. doi:10.1097/01.NNR.0000319495.59746.b8.

Chambers, D. W., Mirchel, R., & Lundergan, W. (2010). An investigation of dentists' and dental students' estimates of diagnostic probabilities. *The Journal of the American Dental Association, 141*(6), 656–666.

Christensen, J. (2010). Accounting errors and errors of accounting. *The Accounting Review, 85*(6), 1827–1838. doi:10.2308/accr.2010.85.6.1827.

Corney, D. (2000). *Designing food with Bayesian belief networks* (pp. 83–94). New York: Springer.

Daland, R., & Pierrehumbert, J. B. (2011). Learning diphone-based segmentation. *Cognitive Science, 35*(1), 119–155. doi:10.1111/j.1551-6709.2010.01160.x.

Data.gov FAQ. (2012). http://www.data.gov/faq/. Accessed 16 June 2012.

Dehghani, M., Iliev, R., & Kaufmann, S. (2012). Causal explanation and fact mutability in counterfactual reasoning. *Mind & Language, 27*(1), 55–85. doi:10.1111/j.1468-0017.2011.01435.x.

DiMaggio, C., Galea, S., & Emch, M. (2010). Spatial proximity and the risk of psychopathology after a terrorist attack. *Psychiatry Research, 176*(1), 55–61. doi:10.1016/j.psychres.2008.10.035.

DiMaggio, C., Galea, S., & Vlahov, D. (2009). Bayesian hierarchical spatial modeling of substance abuse patterns following a mass Trauma: The role of time and place. *Substance Use & Misuse, 44*(12), 1725–1743. doi:10.3109/10826080902963399.

DiMaggio, C., & Li, G. H. (2012). Roadway characteristics and pediatric pedestrian injury. *Epidemiologic Reviews, 34*(1), 46–56. doi:10.1093/epirev/mxr021.

Donaldson, T., & Wollert, R. (2008). A mathematical proof and example that Bayes's theorem is fundamental to actuarial estimates of sexual recidivism risk. *Sexual Abuse-a Journal of Research and Treatment, 20*(2), 206–217. doi:10.1177/1079063208317734.

Droguett, E. L., & Mosleh, A. (2008). Bayesian methodology for model uncertainty using model performance data. *Risk Analysis, 28*(5), 1457–1476. doi:10.1111/j.1539-6924.2008.01117.x.

Elliott, M. N., Fremont, A., Morrison, P. A., Pantoja, P., & Lurie, N. (2008). A new method for estimating race/ethnicity and associated disparities where administrative records lack self-reported race/ethnicity. *Health Services Research, 43*(5), 1722–1736. doi:10.1111/j.1475-6773.2008.00854.x.

Fabianczyk, K. (2011). Decision making on ambiguous stimuli such as prosody by subjects suffering from paranoid schizophrenia, alcohol dependence, and without psychiatric diagnosis. *British Journal of Mathematical and Statistical Psychology, 64*(1), 53–68. doi:10.1348/000711010x492366.

Fienberg, S. E. (2008). The early statistical years. 1947–1967: A conversation with Howard Raiffa. *Statistical Science, 23*(1), 136–149. doi:10.1214/088342307000000104 [Institute of Mathematical Statistics].

Fienberg, S. E. (2011). Bayesian models and methods in public policy and government settings. http://arxiv.org/pdf/1108.2177.pdf. Accessed 26 June 2012.

Fisher, C. R., & Wolfe, C. R. (2011). Assessing semantic coherence in conditional probability estimates. *Behavior Research Methods, 43*(4), 999–1002. doi:10.3758/s13428-011-0099-3.

Foxcroft, D. R., Kypri, K., & Simonite, V. (2009). Bayes' theorem to estimate population prevalence from alcohol use disorders identification test (AUDIT) scores. *Addiction, 104*(7), 1132–1137. doi:10.1111/j.1360-0443.2009.02574.x.

Gomben, P., Lilieholm, R., & Gonzalez-Guillen, M. (2012). Impact of demographic trends on future development patterns and the loss of open space in the California Mojave desert. *Environmental Management, 49*(2), 305–324. doi:10.1007/s00267-011-9749-6.

Grant, T. (2007). Quantifying evidence in forensic authorship analysis. *International Journal of Speech Language and the Law, 14*(1), 1–25. doi:10.1558/ijsll.v14i1.1.

Gregoriades, A., & Sutcliffe, A. (2008a). A socio-technical approach to business process simulation. *Decision Support Systems, 45*(4), 1017–1030. doi:10.1016/j.dss.2008.04.003.

Gregoriades, A., & Sutcliffe, A. (2008b). Workload prediction for improved design and reliability of complex systems. *Reliability Engineering and System Safety, 93*(4), 530–549. doi:10.1016/j.ress.2007.02.001.

Gregoriades, A., Sutcliffe, A., Papageorgiou, G., & Louvieris, P. (2010). Human-centered safety analysis of prospective road designs. *IEEE Transactions on Systems Man and Cybernetics Part A-Systems and Humans, 40*(2), 236–250. doi:10.1109/tsmca.2009.2037011.

Grove, W. M., & Vrieze, S. I. (2009). An exploration of the base rate scores of the Millon clinical multiaxial inventory-III. *Psychological Assessment, 21*(1), 57–67. doi:10.1037/a0014471.

Gunay, H. (2008). The role of externalities and information aggregation in market collapse. *Economic Theory, 35*(2), 367–379. doi:10.1007/s00199-006-0158-7.

Gustafson, L., Klotins, K., Tomlinson, S., Karreman, G., Cameron, A., Wagner, B., et al. (2010). Combining surveillance and expert evidence of viral hemorrhagic septicemia freedom: A decision science approach. *Preventive Veterinary Medicine, 94*(1–2), 140–153. doi:10.1016/j.prevetmed.2009.11.021.

Gutierrez, B. T., Plant, N. G., & Thieler, E. R. (2011). A Bayesian network to predict vulnerability to sea-level rise: Data report. U.S. Geological Survey Data Series 601.

Harris, A. J. L., & Hahn, U. (2009). Bayesian rationality in evaluating multiple testimonies: Incorporating the role of coherence. *Journal of Experimental Psychology-Learning Memory and Cognition, 35*(5), 1366–1373. doi:10.1037/a0016567.

Heuze, Y., & Braga, J. (2008). Application of non-adult Bayesian dental age assessment methods to skeletal remains: The Spitalfields collection. *Journal of Archaeological Science, 35*(2), 368–375. doi:10.1016/j.jas.2007.04.003.

Hoffman, B., & Schraw, G. (2010). Conceptions of efficiency: Applications in learning and problem solving. *Educational Psychologist, 45*(1), 1–14. doi:10.1080/00461520903213618.

Hood, G. M., Barry, S. C., & Martin, P. A. J. (2009). Alternative methods for computing the sensitivity of complex surveillance systems. *Risk Analysis, 29*(12), 1686–1698. doi:10.1111/j.1539-6924.2009.01323.x.

Hossain, M., & Muromachi, Y. (2012). A Bayesian network based framework for real-time crash prediction on the basic freeway segments of urban expressways. *Accident Analysis and Prevention, 45*, 373–381. doi:10.1016/j.aap.2011.08.004.

Hsu, C. I., Shih, M. L., Huang, B. W., Lin, B. Y., & Lin, C. N. (2009). Predicting tourism loyalty using an integrated Bayesian network mechanism. *Expert Systems with Applications, 36*(9), 11760–11763. doi:10.1016/j.eswa.2009.04.010.

http://patft.uspto.gov/netahtml/PTO/index.html. Accessed 14 September 2016.

Inman, D., Blind, M., Ribarova, I., Krause, A., Roosenschoon, O., Kassahun, A., et al. (2011). Perceived effectiveness of environmental decision support systems in participatory planning: Evidence from small groups of end-users. *Environmental Modelling and Software, 26*(3), 302–309. doi:10.1016/j.envsoft.2010.08.005.

Jalen, L., & Mamon, R. (2009). Valuation of contingent claims with mortality and interest rate risks. *Mathematical and Computer Modelling, 49*(9–10), 1893–1904. doi:10.1016/j.mcm.2008.10.014.

Joseph, S. A., Adams, B. J., & McCabe, B. (2010). Methodology for Bayesian belief network development to facilitate compliance with water quality regulations. *Journal of Infrastructure Systems, 16*(1), 58–65. doi:10.1061/(asce)1076-0342(2010)16:1(58).

Joyce, J. (2008). *"Bayes' theorem", The Stanford encyclopedia of philosophy* (Fall 2008 Edition). In E.N. Zalta (Ed.), Retrieved from http://plato.stanford.edu/archives/fall2008/entries/bayes-theorem/.

Kasess, C. H., Stephan, K. E., Weissenbacher, A., Pezawas, L., Moser, E., & Windischberger, C. (2010). Multi-subject analyses with dynamic causal modeling. *NeuroImage, 49*(4), 3065–3074. doi:10.1016/j.neuroimage.2009.11.037.

Kirkos, E., Spathis, C., & Manolopoulos, Y. (2007). Data mining techniques for the detection of fraudulent financial statements. *Expert Systems with Applications, 32*(4), 995–1003. doi:10.1016/j.eswa.2006.02.016.

Kisioglu, P., & Topcu, Y. I. (2011). Applying Bayesian belief network approach to customer churn analysis: A case study on the telecom industry of Turkey. *Expert Systems with Applications, 38* (6), 7151–7157. doi:10.1016/j.eswa.2010.12.045.

Kleiter, G. D. (2007). Implications of natural sampling in base-rate tasks. *The Behavioral and Brain Sciences, 30*(3), 270–271. doi:10.1017/s0140525x07001793 [Editorial Material].

Koelle, D., Pfautz, J., Farry, M., Cox, Z., Catto, G., & Campolongo, J. (2006). Applications of Bayesian belief networks in social network analysis. Charles River Analytics Inc. *22nd Annual Conference on Uncertainty in Artificial Intelligence: UAI '06*, Cambridge, MA.

Kragt, M. E., Newham, L. T. H., Bennett, J., & Jakeman, A. J. (2011). An integrated approach to linking economic valuation and catchment modelling. *Environmental Modelling and Software, 26*(1), 92–102. doi:10.1016/j.envsoft.2010.04.002.

Kusumastuti, D., Hannes, E., Depaire, B., Vanhoof, K., Janssens, D., Wets, G., et al. (2011). An interactive computer-based interface to support the discovery of individuals' mental representations and preferences in decisions problems: An application to travel. *Computers in Human Behavior, 27*(2), 997–1011. doi:10.1016/j.chb.2010.12.004.

Laming, D. (2007). Ordinary people do not ignore base rates. *The Behavioral and Brain Sciences, 30*(3), 272–274. doi:10.1017/s0140525x0700181 [Editorial Material].

Lee, K. J., & Chang, W. (2009). Bayesian belief network for box-office performance: A case study on Korean movies. *Expert Systems with Applications, 36*(1), 280–291. doi:10.1016/j.eswa.2007.09.042.

Lehmann, E. L. (1995). Neyman's statistical philosophy. *Probability and Mathematical Statistics, 15*, 29–36.

Levontin, P., Kulmala, S., Haapasaari, P., & Kuikka, S. (2011). Integration of biological, economic, and sociological knowledge by Bayesian belief networks: The interdisciplinary evaluation of potential management plans for Baltic salmon. *ICES Journal of Marine Science, 68*(3), 632–638. doi:10.1093/icesjms/fsr004.

Linders, B. (2009). Building process improvement business cases using Bayesian belief networks and Monte Carlo simulation. *Software Engineering Institute, CMU/SEI-2009-TN-017.*

Lindgaard, G., Pyper, C., Frize, M., & Walker, R. (2009). Does Bayes have it? Decision support systems in diagnostic medicine. *International Journal of Industrial Ergonomics, 39*(3), 524–532. doi:10.1016/j.ergon.2008.10.011.

Little, T. T., Miller, G., & Guilmette, R. (2003). Internal dosimetry of uranium isotopes using Bayesian inference methods. *Radiation Protection Dosimetry, 105*(1–4), 413–416.

Liu, Q., Otter, T., & Allenby, G. M. (2007). Investigating endogeneity bias in marketing. *Marketing Science, 26*(5), 642–650. doi:10.1287/mksc.1060.0256.

Louvieris, P., Gregoriades, A., & Garn, W. (2010). Assessing critical success factors for military decision support. *Expert Systems with Applications, 37*(12), 8229–8241. doi:10.1016/j.eswa.2010.05.062.

Loyd, K. A. T., & DeVore, J. L. (2010). An evaluation of Feral cat management options using a decision analysis network. *Ecology and Society, 15*(4):10. http://www.ecologyandsociety.org/vol15/iss4/art10/.

Luu, V. T., Kim, S. Y., Nguyen, V. T., & Ogunlana, S. O. (2009). Quantifying schedule risk in construction projects using Bayesian belief networks. *International Journal of Project Management, 27*(1), 39–50. doi:10.1016/j.ijproman.2008.03.003.

Ma, L., Arentze, T., Borgers, A., & Timmermans, H. (2007). Modelling land-use decisions under conditions of uncertainty. *Computers Environment and Urban Systems, 31*(4), 461–476. doi:10.1016/j.compenvurbsys.2007.02.002.

Ma, J. M., Kockelman, K. M., & Damien, P. (2008). A multivariate poisson-lognormal regression model for prediction of crash counts by severity, using Bayesian methods. *Accident Analysis and Prevention, 40*(3), 964–975. doi:10.1016/j.aap.2007.11.002.

Macauley, M., & Laxminarayan, R. (2010). The value of information: 'Methodological frontiers and new applications for realizing social benefit' workshop. *Space Policy, 26*(4), 249–251. doi:10.1016/j.spacepol.2010.08.007.

Mannor, S., & Shinikin, N. (2008). Regret minimization in repeated matrix games with variable stage duration. *Games and Economic Behavior, 63*(1), 227–258. doi:10.1016/j.geb.2007.07.006.

Marhavilas, P. K., Koulouriotis, D., & Gemeni, V. (2011). Risk analysis and assessment methodologies in the work sites: On a review, classification and comparative study of the scientific literature of the period 2000–2009. *Journal of Loss Prevention in the Process Industries, 24*(5), 477–523. doi:10.1016/j.jlp.2011.03.004 [Review].

Martin, J. E., Rivas, T., Matias, J. M., Taboada, J., & Arguelles, A. (2009). A Bayesian network analysis of workplace accidents caused by falls from a height. *Safety Science, 47*(2), 206–214. doi:10.1016/j.ssci.2008.03.004.

Martinez, E. Z., Ruffino-Netto, A., Achcar, J. A., & Aragon, D. C. (2008). Bayesian model for the risk of tuberculosis infection for studies with individuals lost to follow-up. *Revista De Saude Publica, 42*(6), 999–1004.

Mast, B. D. (2010). Measuring neighborhood quality with survey data: A Bayesian approach. *Cityscape: A Journal of Policy Development and Research, 12*(3). 123–143.

Mast, B. D. (2012). Comparing public housing and housing voucher tenants with Bayesian propensity scores. *Cityscape, 14*(1), 55–72.

Mazzocco, T., & Hussain, A. (2012). Novel logistic regression models to aid the diagnosis of dementia. *Expert Systems with Applications, 39*(3), 3356–3361. doi:10.1016/j.eswa.2011.09.023.

McCabe, B., Loughlin, C., Munteanu, R., Tucker, S., & Lam, A. (2008). Individual safety and health outcomes in the construction industry. *Canadian Journal of Civil Engineering, 35*(12), 1455–1467. doi:10.1139/l08-091.

McCarron, C. E., Pullenayegum, E. M., Marshall, D. A., Goeree, R., & Tarride, J. E. (2009). Handling uncertainty in economic evaluations of patient level data: A review of the use of Bayesian methods to inform health technology assessments. *International Journal of Technology Assessment in Health Care, 25*(4), 546–554. doi:10.1017/s0266462309990316 [Review].

McCloskey, J. T., Lilieholm, R. J., & Cronan, C. (2011). Using Bayesian belief networks to identify potential compatibilities and conflicts between development and landscape conservation. *Landscape and Urban Planning, 101*(2), 190–203. doi:10.1016/j.landurbplan.2011.02.011.

McGrayne, S. B. (2011). *The theory that would not die: How Bayes' rule cracked the Enigma code, hunted down Russian submarines, and emerged triumphant from two centuries of controversy.* New Haven: Yale University Press.

McKaughan, D. C., Heath, Z., & McClain, J. T. (2011). *Using a text analysis and categorization tool to generate Bayesian belief networks for use in cognitive social simulation from a document corpus.* Paper presented at the Proceedings of the 2011 Military Modeling & Simulation Symposium, Boston, MA

Michaeli, R., & Simon, L. (2008). An illustration of Bayes' theorem and its use as a decision-making aid for competitive intelligence and marketing analysts. *European Journal of Marketing, 42*(7–8), 804–813. doi:10.1108/03090560810877169.

Milgrom, J., Mendelsohn, J., & Gemmill, A. W. (2011). Does postnatal depression screening work? Throwing out the bathwater, keeping the baby. *Journal of Affective Disorders, 132*(3), 301–310. doi:10.1016/j.jad.2010.09.031 [Review].

Mohaghegh, Z., Kazemi, R., & Mosleh, A. (2009). Incorporating organizational factors into probabilistic risk assessment (PRA) of complex socio-technical systems: A hybrid technique formalization. *Reliability Engineering and System Safety, 94*(5), 1000–1018. doi:10.1016/j.ress.2008.11.006.

Murray-Tuite, P. M., & Fei, X. A. (2010). A methodology for assessing transportation network terrorism risk with attacker and defender interactions. *Computer-Aided Civil and Infrastructure Engineering, 25*(6), 396–410. doi:10.1111/j.1467-8667.2010.00655.x.

Murungweni, C., van Wijk, M. T., Andersson, J. A., Smaling, E. M. A., & Giller, K. E. (2011). Application of fuzzy cognitive mapping in livelihood vulnerability analysis. *Ecology and Society, 16*(4). doi:10.5751/es-04393-160408.

Nasoz, F., Lisetti, C. L., & Vasilakos, A. V. (2010). Affectively intelligent and adaptive car interfaces. *Information Sciences, 180*(20), 3817–3836. doi:10.1016/j.ins.2010.06.034.

Neil, M., Fenton, N., Forey, S., & Harris, R. (2001). Using Bayesian belief networks to predict the reliability of military vehicles. *Computing & Control Engineering Journal, 12*(1), 11–20. doi:10.1049/ccej:20010103.

Neil, M., Hager, D., & Andersen, L. B. (2009). Modeling operational risk in financial institutions using hybrid dynamic Bayesian networks. *Journal of Operational Risk, 4*(1), 3–33.

Newton, A. C., Cayuela, L., Echeverria, C., Armesto, J. J., Del Castillo, R. F., Golicher, D., et al. (2009). Toward integrated analysis of human impacts on forest biodiversity: Lessons from Latin America. *Ecology and Society, 14*(2), 2. http://www.ecologyandsociety.org/vol14/iss2/art2/.

Newton, A. C., Marshall, E., Schreckenberg, K., Golicher, D., Te Velde, D. W., Edouard, F., et al. (2006). Use of a Bayesian belief network to predict the impacts of commercializing non-timber forest products on livelihoods. *Ecology and Society, 11*(2), 24. http://www.ecologyandsociety.org/vol11/iss2/art24/.

Ngai, E. W. T., Hu, Y., Wong, Y. H., Chen, Y. J., & Sun, X. (2011). The application of data mining techniques in financial fraud detection: A classification framework and an academic review of literature. *Decision Support Systems, 50*(3), 559–569. doi:10.1016/j.dss.2010.08.006.

Norrington, L., Quigley, J. L., Russell, A. H., & Van Der Meer, R. B. (2008). Modelling the reliability of search and rescue operations with Bayesian belief networks. *Reliability Engineering and System Safety, 93*(7), 940–949. doi:10.1016/j.ress.2007.03.006.

Oakil, A. M., Ettema, D., Arentze, T., & Timmermans, H. (2011). Longitudinal model of longer-term mobility decisions: Framework and first empirical tests. *Journal of Urban Planning and Development-ASCE, 137*(3), 220–229. doi:10.1061/(asce)up.1943-5444.0000066.

Oaksford, M., & Chater, N. (2009). Precis of Bayesian rationality: The probabilistic approach to human reasoning. *The Behavioral and Brain Sciences, 32*(1), 69. doi:10.1017/s0140525x09000284 [Review].

Padilla, C., Lalloue, B., Zmirou-Navier, D., & Severine, D. (2011). Infant mortality, deprivation and proximity to polluting industrial facilities – a small-scale spatial analysis with census data (Lille metropolitan area, France). *Environment Risques & Sante, 10*(3), 216–221. doi:10.1684/ers.2011.0455.

Pearl, J. (1985). A model of self-activated memory for evidential reasoning. In *7th Conference of the Cognitive Science Society, University of California, Irvine, August 1985* (pp. 329–334).

Peterson, L. T., Ford, E. W., Eberhardt, J., Huerta, T. R., & Menachemi, N. (2011). Assessing differences between physicians' realized and anticipated gains from electronic health record adoption. *Journal of Medical Systems, 35*(2), 151–161. doi:10.1007/s10916-009-9352-z.

Raiffa, H. (2012). Howard Raiffa on decision trees. http://www.hbs.edu/centennial/im/inquiry/sections/3/b/page15.html. Accessed 26 June 2012.

Ranta, J., Matjushin, D., Virtanen, T., Kuusi, M., Viljugrein, H., Hofshagen, M., et al. (2011). Bayesian temporal source attribution of foodborne zoonoses: Campylobacter in Finland and Norway. *Risk Analysis, 31*(7), 1156–1171. doi:10.1111/j.1539-6924.2010.01558.x.

Replogle, W. H., Johnson, W. D., & Hoover, K. W. (2009). Using evidence to determine diagnostic test efficacy. *Worldviews on Evidence-Based Nursing, 6*(2), 87–92.

Rupp, J. D., Flannagan, C. A. C., Hoff, C. N., & Cunningham, R. M. (2010). Effects of osteoporosis on AIS 3 + injury risk in motor-vehicle crashes. *Accident Analysis and Prevention, 42*(6), 2140–2143. doi:10.1016/j.aap.2010.07.005.

Saravanan, V. S. (2010). Negotiating participatory irrigation management in the Indian Himalayas. *Agricultural Water Management, 97*(5), 651–658. doi:10.1016/j.agwat.2009.12.003.

Scarlata, S., Pedone, C., Conte, M. E., & Incalzi, R. A. (2009). Accuracy of spirometry in diagnosing pulmonary restriction in elderly people. *Journal of the American Geriatrics Society, 57*(11), 2107–2111. doi:10.1111/j.1532-5415.2009.02525.x.

Schneider, I. (2005). Abraham De Moivre, the doctrine of chances (1718, 1738, 1756). In I. Grattan-Guinness (Ed.), *Landmark writings in western mathematics 1640–1940*. Amsterdam: Elsevier.

Schultz, G., Thurgood, D., Olsen, A., & Reese, C. S. (2011). Analyzing raised median safety impacts using Bayesian methods. Presented at the 90th meeting of the Transportation Research Board, Washington, DC.

Sebastiani, P., Ramoni, M., Cohen, P., Warwick, J., & Davis, J. (1999). Discovering dynamics using Bayesian clustering. In D. Hand, J. Kok, & M. Berthold (Eds.), *Advances in intelligent data analysis*. Lecture notes in computer science (Vol. 1642, pp. 199–209). Berlin: Springer.

Sevcikova, H., Raftery, A. E., & Waddell, P. A. (2007). Assessing uncertainty in urban simulations using Bayesian melding. *Transportation Research Part B-Methodological, 41*(6), 652–669. doi:10.1016/j.trb.2006.11.001.

Small, M. J. (2008). Methods for assessing uncertainty in fundamental assumptions and associated models for cancer risk assessment. *Risk Analysis, 28*(5), 1289–1307. doi:10.1111/j.1539-6924.2008.01134.x.

Smid, J. H., Swart, A. N., Havelaar, A. H., & Pielaat, A. (2011). A practical framework for the construction of a biotracing model: Application to Salmonella in the pork slaughter chain. *Risk Analysis, 31*(9), 1434–1450. doi:10.1111/j.1539-6924.2011.01591.x.

Smith, C., Felderhof, L., & Bosch, O. J. H. (2007). Adaptive management: Making it happen through participatory systems analysis. *Systems Research and Behavioral Science, 24*(6), 567–587. doi:10.1002/sres.835.

Social Sciences Citation Index®, accessed via Web of Science® (2012). Thomson Reuters. Accessed June 1, 2012. http://thomsonreuters.com/products_services/science/science_products/a-z/social_sciences_citation_index/

Staker, R. J. (1999). Use of Bayesian belief networks in the analysis of information system network risk. *Information, Decision and Control, 1999. IDC 99. Proceedings. 1999* (pp. 145–150). doi:10.1109/idc.1999.754143.

Starr, C., & Shi, P. (2004). An introduction to Bayesian belief networks and their applications to land operations. *Network, 391*(6670). doi:10.1038/36103.

Stelzenmuller, V., Schulze, T., Fock, H. O., & Berkenhagen, J. (2011). Integrated modelling tools to support risk-based decision-making in marine spatial management. *Marine Ecology Progress Series, 441*, 197–212. doi:10.3354/meps09354.

Swiler, L. P. (2006). Bayesian methods in engineering design problems. *Optimization, SAND* 2005-3294.

Tang, D. W., Yang, J. B., Chin, K. S., Wong, Z. S. Y., & Liu, X. B. (2011). A methodology to generate a belief rule base for customer perception risk analysis in new product development. *Expert Systems with Applications, 38*(5), 5373–5383. doi:10.1016/j.eswa.2010.10.018.

Thodi, P. N., Khan, F. I., & Haddara, M. R. (2010). The development of posterior probability models in risk-based integrity modeling. *Risk Analysis, 30*(3), 400–420. doi:10.1111/j.1539-6924.2009.01352.x.

Vans, O. (1998). A belief network approach to optimization and parameter estimation: Application to resource and environmental management. *Artificial Intelligence, 101*(1–2), 135–163. doi:10.1016/s0004-3702(98)00010-1.

Verma, M. (2011). Railroad transportation of dangerous goods: A conditional exposure approach to minimize transport risk. *Transportation Research Part C-Emerging Technologies, 19*(5), 790–802. doi:10.1016/j.trc.2010.07.003.

Walshe, T., & Burgman, M. (2010). A framework for assessing and managing risks posed by emerging diseases. *Risk Analysis, 30*(2), 236–249. doi:10.1111/j.1539-6924.2009.01305.x.

Wollert, R. (2007). Poor diagnostic reliability, the null-Bayes logic model, and their implications for sexually violent predator evaluations. *Psychology, Public Policy, and Law, 13*(3), 167–203. doi:10.1037/1076-8971.13.3.167.

Wong, W., Liu, W., & Bennamoun, M. (2009). A probabilistic framework for automatic term recognition. *Intelligent Data Analysis, 13*(4), 499–539. doi:10.3233/ida-2009-0379.

Zellner, A. (2002). Information processing and Bayesian analysis. *Journal of Econometrics, 107*(1–2), 41–50. doi:10.1016/s0304-4076(01)00112-9.

Zimmerman, L. A., Sestokas, J. M., Burns, C. A., Grover, J., Topaz, D., & Bell, J. (2010). *Improving attrition rates in the M1A1/M1A2 Master Gunner course* (Final report prepared under Subcontract ARI-ARA-07-001, Task order 14524 (DRC); Contract # W74V8H-04-D-0048 for U.S. Army Research Institute for the Behavioral and Social Sciences, Fort Knox, KY). Fairborn, OH: Klein Associates Division of Applied Research Associates.

Zorrilla, P., Carmona, G., De la Hera, A., Varela-Ortega, C., Martinez-Santos, P., Bromley, J., et al. (2010). Evaluation of Bayesian networks in participatory water resources management, Upper Guadiana Basin, Spain. *Ecology and Society, 15*(3), 12. http://www.ecologyandsociety.org/vol15/iss3/art12/.

Chapter 3
Statistical Properties of Bayes' Theorem

This chapter provides an introduction to statistical terminology and definitions, and a review of the basic statistical properties associated with modeling Bayesian Belief Networks (BBN).

3.1 Axioms of Probability

The following are three basic axioms of probability that underpin the concepts of Bayes' theorem and BBN[1]:

- Chances are always at least zero (never negative), $P(A) \geq 0$,
- the chance that something happens in a universe is always 100 %, $P(\cup) = 1$, and
- if two events cannot both occur at the same time (if they are disjoint or mutually exclusive), the chance that either one occurs is the sum of the chances that each occurs, $P(A \cup B) = P(A) + P(B)$. For non-mutually exclusive events, this is $P(A \cup B) = P(A) + P(B) - P(A \cap B)$.

From these axioms, we can derive other mathematical facts about probabilities.

3.2 Base-Rate Fallacy

The *base-rate fallacy* fails to account for *prior* probabilities (base-rates), when computing *conditionals*, from other *conditional* probabilities. It is related to the Prosecutor's Fallacy. For instance, suppose a test for the presence of some condition has a 1 % chance of a false positive result (the test says the condition is present when

---

[1]Bolstad (2007).

© Springer International Publishing AG 2016
J. Grover, *The Manual of Strategic Economic Decision Making*,
DOI 10.1007/978-3-319-48414-3_3

it is not), and a 1 % chance of a false negative result (the test says the condition is absent when the condition is present), then these results are 99 % accurate. What is the chance that a condition that tests positive actually exist? The intuitive answer is 99 %, but that is not necessarily true. The correct answer depends on the fraction f of items in the population that have the condition (and if the item tested is selected at random from the population). The chance that a randomly selected item tests positive is $0.99 * f / (0.99 * f + 0.01 * (1 - f))$.

3.3 Bayes' Theorem

Equation (3.1) is Bayes' theorem in its simplest form:

$$P(B_i|A) = \frac{P(A, B_i)}{P(A)},\tag{3.1}$$

which is, the *conditional* probability of an unobservable event B_i, given the probability of an observable event A, $P(B_i|A)$, is equal to the *joint* probability of event A and event B_i, $P(A, B_i)$, divided by the probability of the event A, $P(A)$. Equation (3.2) expands it to:

$$P(B_i|A) = \frac{P(A|B_i) * P(B_i)}{P(A)}\tag{3.2}$$

where the numerator of the right side of (3.1) is expanded using the chain rule to the conditional probability of event A given event B_i, $P(A|B_i)$ times the probability of event B_i, $P(B_i)$.

Below are definitions of the theorem's probabilities:

3.3.1 Prior Probability

$P(A)$ is the prior probability, which are either know or unknown, subjective, and we use them as starting values. This parameter is updated through the posterior probability during the Bayesian updating process.

3.3.2 Conditional Probability

Both $P(A|B_i)$ and $P(B_i|A)$ are *conditional* probabilities. In the first, A is conditioned on B_i and the second, B_i is conditioned on A.

3.3.3 Joint and Marginal Probability

$P(A, B_i)$ is the *joint* probability that this event occurs at the same time, on the same iteration, of a random experimental trial.

3.3.4 Posterior Probability

In additional to being a *conditional* probability, here $P(B_i|A)$ is the *posterior* probability. We update this parameter, using the chain rule and propagation.

3.4 Joint and Disjoint Bayesian Belief Network Structures

A BBN is a mathematical model, containing *conditional* and *unconditional* probabilities, of all related events in the respective variables or nodes. The *joint* probabilities are the engine of probability updating. Based on the statistical structure of the BBN, when we invoke one or more nodal events are invoked,[2] then through these *joints*, we revise the *posteriors*. These BBN are a linked series of nodes, joint or disjoint, containing random variables. Each random variable contains a partial subset of the sample space. When we link these nodes, they have a joint and when they are not linked they have a disjoint structure. When linked, these are the paths by which they propagate probabilities. Even when we have non-linked nodes, they still form *joints*. From these *joints*, we can compute the *posteriors*, which revise the *priors*, during this chaining process.

3.4.1 Joint BBN Structure

Joint nodes are dependent of the other nodes in the BBN. *Joint* BBN structures consist of those that are fully-specified, or joined together. Each node is connected through a series of paths.[3] Each element is dependent on the other, i.e., they are not statistically independent. Figure 3.1 is a fully-specified BBN. Note that all paths are fully-specified, i.e., there is a path to each node combination, beginning with the parent node, A: Paths $[A \rightarrow B|A \rightarrow C|B \rightarrow C]$, where we computed the *joints* using the chain rule as: $P(A_1, B_1, C_1) = P(A_1|B_1, C_1) \times P(B_1, C_1) \times P(C_1)$.

---

[2]Meaning when these events occur in nature.
[3]Except in the case of a two-node BBN where there is only one possible path.

Fig. 3.1 Fully specified BBN
paths: $[A \rightarrow B|A \rightarrow C|B \rightarrow C]$

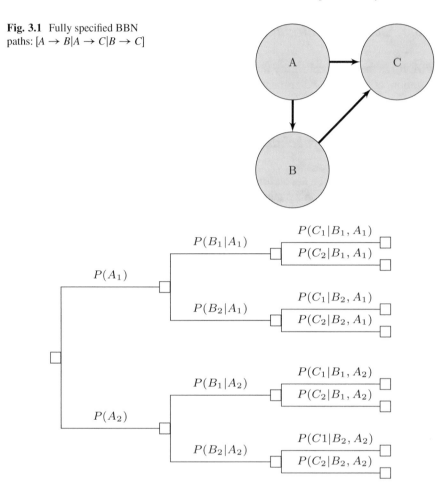

Fig. 3.2 Fully specified BBN conditional probabilities

Figure 3.2 is a tree diagram, highlighting the *conditional* probabilities of the BBN
in Fig. 3.1.

3.4.2 Disjoint (Pairwise) Bayesian Belief Network Structure

Disjoint BBN can contain independent and independent nodes. The BBN in Fig. 3.3
is a partially-specified, and disjoint BBN. It is disjoint because the link between *B*
and *C* is missing; so there is no dependent relationship here. They differ in that the
events of the nodes are statistically independent. For example, node B is independent
of node C.[4]

---

[4]Disjoint BBN are subset of parent joint BBN.

Fig. 3.3 Partially specified
BBN paths: $[A \rightarrow B|[A \rightarrow C]$

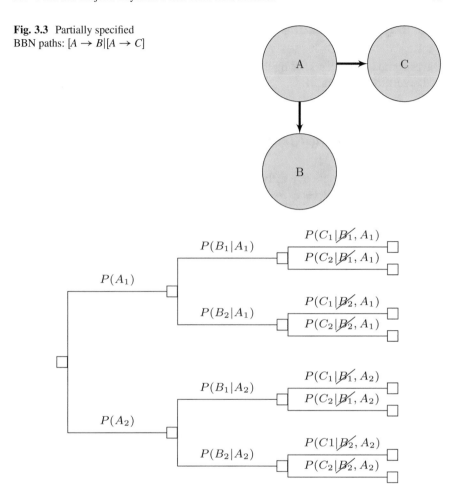

Fig. 3.4 Partially specified BBN conditional probabilities

More explicitly, two sets A and B are disjoint if and only if their intersection $A \cap B$ is the empty set. It follows from this definition, that every set is disjoint from the empty set, and that the empty set $\emptyset$ is the only set that is disjoint from itself. A family of sets is pairwise disjoint or mutually disjoint if every two different sets in the family are disjoint. For example, the collection of sets $\{\{1\}, \{2\}, \{3\}, \ldots\}$ is pairwise disjoint. Figure 3.3 represents disjoint events, where C is disjoint of B.[5]

Figure 3.4 is a tree diagram highlighting the *conditional* probabilities of the BBN in Fig. 3.3. The path between Node B and C is missing, so they are independent of each other, as we designate with the strikethrough. For example, $P(C_1|\cancel{B_1}, A_1) = P(C_1|A_1)$.

---

[5]Taken from:https://en.wikipedia.org/wiki/Disjoint_sets.

3.5   Bayesian Updating

Bayesian updating occurs, using the chain rule, where the *joints* are updated, as we add events to the chain.

3.5.1   Fully Specified Joint BBN

Figure 3.5 is an example of a fully-specified, joint BBN. The joint $P(C_1, B_1, A_1)$ is updated by the joint events $P(B_1, A_1)$ and $P(A_1)$, which is reflected in the *posterior* probability $P(A_1|B_1, C_1)$.

3.5.2   Partially Specified Disjoint BBN

Figure 3.6 is an example of a partially-specified joint BBN. The joint $P(C_1, B_1, A_1)$ is updated by the joint event $P(B_1, A_1)$, which is reflected in the *posterior* probability $P(A_1|B_1, C_1)$.

3.6   Certain Event

An event is certain, if its probability is 100 %. Even if an event is certain, it might not occur. By the complement rule, the chance that it does not occur is 0 %.

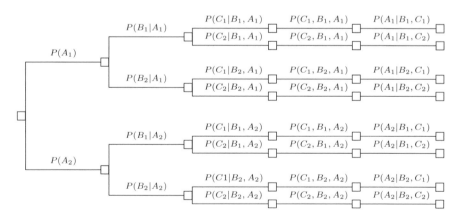

Fig. 3.5 Fully specified BBN conditional probabilities

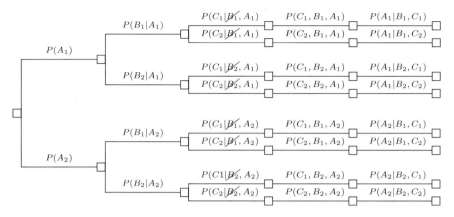

Fig. 3.6 Example of partially specified Bayesian updating

3.7 Categorical Variable

A variable, whose value ranges over categories, such as {*red, green, blue*}, {*male, female*}, {*Arizona, California, Montana, New York*}, {*short, tall*}, {*Asian, African-American, Caucasian, Hispanic, Native American, Polynesian*}, {*straight, curly*}, *etc*. Some categorical variables are ordinal.

3.8 Chain (Product) Rule

The *chain* (or product) rule for probability, for non-zero events, allows for the expansion of members of a set of random variables, from a joint distribution, across any BBN, using only *conditional* probabilities. For example, if we have a collection of events $P(A_{(n)}, ..., A_{(1)}) = P(A_{(n)}|P(A_{(n-1)}, ..., A_{(1)}))$.

Chaining this process across a BBN, creates the product using the chain rule in Eq. (3.3):

$$P(\cap_{k=1}^{n} A_k) = \prod_{k=1}^{n} P(A_k | \cap_{j=1}^{k-1} A_j) \tag{3.3}$$

This is a well-known, general form of the rule. For example: $P(A_1, A_2, A_3) = P(A_1|A_2, A_3) \times P(A_2|A_3)P(A_3)$.[6]

---

[6]We also use the comma to represent the joint symbol $\cap$ for simplicity. For example: $P(A_1 \cap A_2) = P(A_1, A_2)$.

3.9   Collectively Exhaustive

A set of events is *collectively exhaustive*, if one of the events must occur.

3.10   Combinations and Permutations

3.10.1   Combinations

The number, of *combinations* of n things taken k at a time, is the number of ways of picking a subset of k of the n things, without replacement, and without regard to the order in which we select the elements of the subset. The number of such combinations is $_nC_k = n!/[k!(n-k)!]$, where $k!$ (pronounced "k Factorial") is $k \times (k-1) \times (k-2) \, x \ldots x \, 1$. The numbers $_nC_k$ are also called the *Binomial coefficients*. From a set that has n elements, one can form a total of 2^n subsets of all sizes. For example, from the set $a, b, c,$ which has 3 elements, to form the $2^3 = 8$ subsets $\{\}$,[7] $\{a\}, \{b\}, \{c\}, \{a,b\}, \{a,c\}, \{b,c\}, \{a,b,c\}$. Because of the number of subsets with k elements, one can form a set with n elements of $_nC_k$, and the total number of subsets is the sum of all possible subsets of each size. It follows that $_nC_0 + {_nC_1} + {_nC_2} + \ldots + {_nC_n} = 2^n$.

For each synthetic BBN in Chaps. 7–9, we can determine the combinations, using Eq. (3.4):

$$Path_n = \frac{n!}{(n-r)! * (r!)}, \tag{3.4}$$

where $n =$ the number of possible node combinations and repetition where order are not important using, the base nodal combinations of a fully-specified BBN consisting of the nodal combinations (or objects), and $r =$ the number of selected nodal-combination paths.[8]

3.10.2   Permutations

A *permutation* of a set is an ordered arrangement of its elements. If the set has n things in it, there are $N!$ different orderings of its elements. For the first element in an ordering, there are n possible choices. For the second, there remain $n - 1$ possible choices. For the third, there are $n - 2$, etc., and for

---

[7]This is the null set.

[8]We used the online combinations calculator @ https://www.mathsisfun.com/combinatorics/combinations-permutations-calculator.html (14 September 2016).

the nth element of the ordering, there is a single choice remaining. By the fundamental rule of counting, the total number of sequences is n x $(n - 1)$ x $(n - 2)$ x ... x 1. Similarly, the number of orderings of length k one can form from $n \geq k$ things is n x $(n - 1)$ x $(n - 2)$ x ... x $(n - k + 1) = n!/[k!(n - k)!]$. This is denoted $_nP_k$, the number of permutations of n things taken k, at a time.

3.11 Complement and Complement Rule

The *complement* of a subset, is the collection of all elements of the set, that are not elements of the subset. For example, $P(A^c)$ is the *complement* of $P(A)$. The *complement rule* states that the probability of the complement of an event is 100 % minus the probability of the event: $P(A^c) = 100\% - P(A)$.

3.12 Conditional and Unconditional Probability

3.12.1 Conditional Probability

Suppose the probability that some event A occurs, and we learn that the event B occurred. How should we update the probability of A to reflect this new knowledge? We do this using the *conditional* probability. It states that the additional knowledge of B occurring, affects the probability that A occurred. Suppose that A and B are mutually exclusive. If B occurred, A did not, so the *conditional* probability that A occurred given that B occurred is zero. At the other extreme, suppose that B is a subset of A, so that A must occur whenever B does. If we learn that B occurred, A must have occurred, so the *conditional* probability that A occurred, given that B occurred, is 100 %. For in-between cases, where A and B intersect, but B is not a subset of A, the *conditional* probability of A given B is a number between zero and one. Basically, one "restricts" the outcome space S, to consider only the part of S, that is in B, because we know that B occurred. For A to have happened, given that B happened, requires that A, B happened, so we are interested in the event A, B. To have a legitimate probability, requires that $P(S) = 100\%$, so if we are restricting the outcome space to B, we need to divide by the probability of B, to make the probability of this new S be 100 %. On this scale, the probability that A, B happened is $P(A, B)/P(B)$. This is the definition of the *conditional* probability of A given B, provided $P(B)$ is not zero (division by zero is undefined).[9] *Conditional* probabilities satisfy the axioms of probability, just as ordinary probabilities do.

---

[9]Note that the special cases $A, B =$ (A and B are mutually exclusive) and $A, B = B$ (B is a subset of A) agree with our intuition, as described at the top of this paragraph.

3.12.2   Unconditional Probability

When the event *A* is invoked, then the *unconditional probability* of *B* given *A* is the unconditional probability of that part of *B*, that is also in *A*. When we multiply it by a scale factor $1/P(A)$ it becomes a conditional probability. For example, $P(B|A) = P(B)$ (Bolstad 2007).

3.13   Counting and Countable Set and Uncountable Set

3.13.1   Counting

To *count* a set of things is to put it in one to one correspondence, with a consecutive subset of the positive integers starting with 1.

3.13.2   Countable and Countable Set

A *Countable set* occur when we can place its elements, in one-to-one correspondence, with a subset of the integers. For example, the sets {0, 1, 7, −3}, {*red, green, blue*}, {. . . , −2, −1, 0, 1, 2, . . .}, {*straight, curly*}, and the set of all fractions are countable. An *uncountable set* is not countable. The set of all real numbers is uncountable.

3.14   Complement and Complement Rule

The *complement* of a subset is the collection of all elements of the set, that are not elements of the subset. The *complement rule* states that the probability of the complement of an event is 100 % minus the probability of the event: $P(A^c) = 100\% - P(A)$.

3.15   Discrete Variable

A *discrete variable* is quantitative, whose set of possible values is countable. Typical examples of discrete variables are a subset of the integers, such as Social Security numbers, the number of people in a family, ages rounded to the nearest year, etc. Discrete variables are "chunky." A discrete random variable is one, whose set of possible values is countable. A random variable is discrete, if and only if, its cumulative probability distribution function is a stair-step function; i.e., if it is piecewise constant and only increases by jumps.

3.16 Disjoint or Mutually Exclusive Events (Sets)

Two events are *disjoint or mutually exclusive*, if the occurrence of one is incompatible with the occurrence of the other; that is, if they cannot both happen at once (if they have no outcome in common). Equivalently, two events are disjoint, if their intersection is the empty set. Two sets are disjoint, or mutually exclusive, if they have no element in common. Equivalently, two sets are disjoint, if their intersection is the empty set.

3.17 Event

An event is a subset of outcome space. An event determined by a random variable takes the form $A = (X$ *is in* $A)$. When the random variable X is observed, this determines if A occurs: if the value of X happens to be in A, A occurs; if not, A does not occur. Joint events contain two, or more characteristics.

Independent. Two events A and B are (statistically) independent, if the chance that they both happen simultaneously is the product of the chances that each occurs individually; i.e., if $P(A, B) = P(A)P(B)$. Learning that one event occurred does signal that the other event has occurred. The conditional probability of A given B is the same as the unconditional probability of A, i.e., $P(A|B) = P(A)$. Two random variables, X and Y, are independent, if all events they determine are independent. For example, if the event $a < X \leq b$ is independent of the event $c < Y \leq d$ for all choices of a, b, c, and d, then a collection of more than two random variables is independent, if for each event in the subset it is independent of events in the complement of the subset. For example, the three random variables X, Y, and Z are independent, if every event determined by X is independent of every event determined by Y and every event determined by X is independent of every event determined by Y and Z, and every event determined by Y is independent of every event determined by X and Z, and every event determined by Z is independent of every event determined by X and Y.

3.18   Factorial

For an integer k that is greater than or equal to 1, $k!$ (pronounced "k factorial") is $kx(k-1)x(k-2)x\ldots x1$. By convention, $0! = 1$. There are $k!$ ways of ordering k distinct objects. For example, $9!$ is the number of batting orders of 9 baseball players, and $52!$ is the number of different ways one can order a standard deck of playing cards.

3.19   Intersection and Union (of Sets)

3.19.1   Intersection

The *intersection* of two or more sets is the common shared elements of these sets, the elements contained in every one of the sets. The intersection of the events A and B is written, "$A \cap B$."

3.19.2   Union

The *union* of two or more sets is the set of objects contained by at least one of the sets. The union of the events A and B is written as, "$A \cup B$."

3.20   Independence and Pairwise Independence

3.20.1   Independence

If $P(A|B) = P(A)$ or $P(B|A) = P(B)$, then the two events are independent. In other words, if events A and B are independent then $P(A, B) = P(A) \times P(B)$. If they are independent then there is not a cause and effect relationship between the events; i.e., event A does not cause event B.

3.20.2   Pairwise Independence

Independence can be extended to more than two events. For example, events A, B and C are independent if: A and B are independent; A and C are independent and B and C are independent (pairwise independence), i.e., $P(A, B, C) = P(A) \times P(B) \times P(C)$.[10]

---

[10]See: http://www.stats.gla.ac.uk/steps/glossary/probability.html#event.

3.21   Joint and Marginal Probability Distribution and Marginalization

3.21.1   Joint Probability Distribution

If X_1, X_2, ...,X_k are random variables, defined for the same experiment, their *joint* probability distribution gives the probability of events determined by the collection of random variables: for any collection of sets $\{A_1, \ldots, A_k\}$, the joint probability distribution determines $P(X_1$ is in $A_1)$ and $(X_2$ is in $A_2)$ and ... and $(X_k$ is in $A_k)$.

3.21.2   Marginal Probability Distribution

The *marginal* probability distribution of a set of random variables with a joint probability distribution is the distribution of each variable regardless of the values of the remaining variables. One can find the marginal distribution of a discrete random variable X_1 that has a joint distribution with other discrete random variables from the joint distribution, by summing over all possible values of the other variables. Equation (3.5) computes the *marginal* probability of event A:

$$P(A) = P(A + B_1)P(A + B_2) + \ldots + P(A + B_k) \tag{3.5}$$

where $B_1, B_2, + \ldots + B_k$ are k mutually exclusive and collectively exhaustive events.

3.21.3   Marginalization

Marginalization is a technique when dealing with an unknown parameter and *conditional* probabilities, using the law of total probability (Sect. 3.22).[11,12] Using a three event example:

$$P(A_i|B, C) = \frac{P(A_i|C) \times P(B|A_i, C)}{P(B|C)} \tag{3.6}$$

We can update our belief in hypothesis A_i given the additional evidence B and the background information C. The left-hand term, $P(A_i|B, C)$ is the *posterior* probability or the probability of A_i after an analyst considers the effect of B given C.

---

[11] We derived this definition from Hebert et al. (2007).

[12] In this manual, strikethroughs represent disjoint events, that are marginalized away, because they are independent of the other events.

The term $P(A_i|C)$ is the *prior* probability of A_i given C alone. The term $P(B|A_i, C)$ is the *likelihood*, and gives the probability of the evidence assuming the hypothesis A, and the background information C is true. Finally, the last term $P(B|C)$ is the expectedness, or how expected the evidence is given in C. It is independent of A_i and an analyst can regard it as a marginalizing or scaling factor.

We can rewrite this as $P(B|C) = \sum_i P(B|A_i, C)P(A_i|C)$ where i denotes a specific hypothesis A_i, and the summation is taken over a set of hypotheses, which are mutually exclusive and collectively exhaustive (their prior probabilities sum to 1).[13]

3.22   Law of Total Probability

The *law of total probability* is the proposition, that if $\{B_n : n = 1, 2, 3, \dots\}$ is a finite or countably infinite partition of a sample space (in other words, a set of pairwise disjoint events whose union is the entire sample space) and each event B_n is measurable, then for any event A of the same probability space:

$P(A) = \sum_n P(A \cap B_n) = \sum_n P(A|B_n) \times P(B_n)$, for any n, for which $P(B_n) = 0$. We omit these terms from the summation because $P(A|B_n)$ is finite. The summation is a weighted average and the marginal probability, $P(A)$, is the "average probability."[14]

3.23   Mean (Arithmetic Mean)

The sum of a list of numbers divided by the total numbers N.

3.24   Ordinal Variable

An *ordinal variable* is one whose possible values have a natural order, such as $\{short, medium, long\}$, $\{cold, warm, hot\}$, $\{0, 1, 2, 3, \dots\}$. In contrast, a variable, whose possible values are $\{straight, curly\}$, $\{Arizona, California, Montana, NewYork\}$ is not naturally ordinal. Arithmetic with the possible values of an *ordinal variable* does not make sense, but saying that "one possible value is larger than another does."

---

[13]Hebert et al. (2007) also suggests that it is important to note that these probabilities are conditional. They specify the degree of belief in some proposition or propositions, based on an initial assumption that some other propositions are true.

[14]See: https://en.wikipedia.org/wiki/Law_of_total_probability.

3.25   Outcome (Outcome Space)

The *outcome* or *outcome space* is the set, of all possible outcomes, of a given random experiment and we denote this by S.

3.26   Parameter

A *parameter* is a numerical property, of a population, such as its mean.

3.27   Partition

A *partition* of an event A is a collection of events $\{A1, A2, A3, \ldots\}$ such that the events in the collection are disjoint and their union is A. That is, $A_j A_k = $ unless $j = k$, and $A = A_1 \cup A_2 \cup A_3 \cup \ldots$. If the event A is not specified, we assume it to be the entire outcome S.

3.28   Population

A *population* is a collection of studied units. Units can be people, places, objects, epochs, drugs, procedures, and many other things. Much of statistics is concerned with estimating numerical properties (parameters) from a random sample of a population.

3.29   Probability and Probability Sample

The *probability*, chance, of an event is a real number between zero and 100 %. The meaning, interpretation, of probability is the subject of theories of probability, which differ in their interpretations. However, any rule for assigning probabilities to events has to satisfy the axioms of probability. A *probability sample* is a draw, from a population using a random mechanism, so that every element of the population has a known chance of ending up in the sample.

3.30 Sample, Sample Size, Sample Space, Random Sample, Simple Random Sample, Random Experiment (Event), and Random Variable

3.30.1 Sample

A *sample* is a collection of units from a population.

3.30.2 Sample Size

A *sample size n* is the number of elements, in a sample, from a population.

3.30.3 Sample Space

In a *sample space*, each trail has as its outcome one of the elements of *S*, which is the set of all possible outcomes of one single trial. *U* contains all possible outcomes. Any subset of the sample space is an event.[15,16]

3.30.4 Random Sample

A *random sample* consists of members chosen at random from a given population, so that one can compute the chance of obtaining any particular sample. *n* is the number of units in the sample size and *N* denotes the population. One can draw random samples with or without replacing objects between draws; drawing all *n* objects in the sample at once (a random sample without replacement), or drawing the objects one at a time, replacing them in the population between draws (a *random sample*, with replacement). In a *random sample* with replacement, any given member of the population can occur in the sample more than once. In a random sample without replacement, any given member of the population can appear only once.

---

[15] In this manual strikethroughs, represent disjoint events that are marginalized away because they are independent of the other events.

[16] See: http://www.stats.gla.ac.uk/steps/glossary/probability.html#event.

3.30.5 Sample Random Sample

A *simple random sample* is sampling at random without replacement (without replacing the units between draws). We can construct a simple random sample of size *n* from a population of $N \geq n$ units by assigning a random number between zero and one to each unit in the population; then taking as the sample, those units that this method assigned the *n* largest random numbers.

3.30.6 Random Experiment (Event)

A *random experiment* is an experiment or trial that has unpredictable short-run frequency of outcomes but has predictable long-run outcomes in repeated trials. Note that "random" is different from "haphazard," which does not necessarily imply long-term regularity.

3.30.7 Random Variable

A *random variable* is an assignment of numbers of possible outcomes in a random experiment. Consider tossing three coins. The number of heads showing, when the coins land, is a random variable: it assigns the number 0 to the outcome $\{T, T, T\}$, the number 1 to the outcome $\{T, T, H\}$, the number 2 to the outcome $\{T, H, H\}$, and the number 3 to the outcome $\{H, H, H\}$.

3.31 Real Number

Real numbers are all numbers that one can represent as fractions (rational numbers), either proper or improper, and all numbers between the rational numbers.

The real numbers comprise the rational numbers, and all limits of Cauchy sequences of rational numbers where the Cauchy sequence is the absolute value metric. (More formally, the real numbers are the completion of the set of rational numbers, in the topology, induced by the absolute value function.) Real numbers contain all integers, fractions, and irrational (and transcendental) numbers, such as π, and $2\frac{1}{2}$. There are unaccountably many real numbers between 0 and 1; in contrast, there are only countably many rational numbers between 0 and 1.

3.32   Set, Subset, Member of a Set, and Empty Set

3.32.1   Set

A *set* is a collection of things without regard to their order.

3.32.2   Subset

A *subset* of a given set is a collection of things that belong to the original set where each element must belong to the original set. Not every element of the original set is in a subset. If this were true, then that subset would be identical to its originating set.

3.32.3   Member (Element) of a Set

Something is a *member (or element) of a set* if it is one of the things in the set.

3.32.4   Empty Set

The *empty set*, denoted {} or Ø, is the set that has no members.

3.33   Theories of Probability

A *theory of probability* is a way of assigning meaning to probability statements, such as "the chance that a thumb tack lands point-up is 2/3." It connects the mathematics of probability, which is the set of consequences of the axioms of probability with the real world of observation and experiment. There are several common theories of probability. According to the frequency theory of probability, the probability of an event is the limit of the percentage of times that the event occurs in repeated, independent trials, under essentially the same circumstances. According to the subjective *theory of probability*, a probability is a number that measures how strongly we believe an event will occur. The number is on a scale of 0–100 %, with 0 % indicating that we are completely sure it will not occur, and 100 %, indicating that we are completely sure that it will occur. According to the theory of equally likely outcomes, if an experiment has n possible outcomes, and (by symmetry) there is no reason that any of these outcomes should occur preferentially to any of the others then the chance of each outcome is $\frac{100\,\%}{n}$. Each of these theories has its limitations, its proponents and its detractors.

3.34   Unit

A *unit* is a member of a population.

3.35   Venn Diagram

A *Venn diagram* shows the relations, among sets or events, using diagrams or pictures. We usually draw the universal set or outcome space as a rectangle where we represent sets as probability regions within this rectangle, and the overlapping regions are the intersection of the sets. If the regions do not overlap, then we say the sets are disjoint or mutually exclusive.

Figure 3.7 is a graphical representation of a *Venn diagram* where the center region $= A \cap B \cap C$.

3.36   The Algebra of Sets

The following are a number of applicable general laws about sets that follow from the definitions of set theoretic operations, subsets, etc.[17] Stoll (1979) offers the following two theorems:

3.36.1   Theorem 1

For any subsets, *A, B, & C* of a set *U*, the following equations are identities[18]:

Fig. 3.7 Example of a Venn diagram

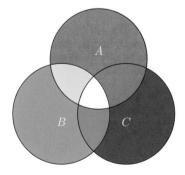

---

[17] See Stoll (1979) for the proofs to these identities.
[18] Here, $\bar{A} = U - A$.

1. $A \cup (B \cup C) = (A \cup B) \cup C$	$A \cap (B \cap C) = (A \cap B) \cap C$	Associative law
2. $A \cup B = B \cup A$	$A \cap B = B \cap A$	Communicative law
3. $A \cup (B \cap C) = (A \cup B) \cap (A \cup C)$	$A \cap (B \cup C) = (A \cap B) \cup (A \cap C)$	Distributive law
4. $A \cup \emptyset = A$	$A \cap \bar{U} = A$	Identity laws
5. $A \cup \bar{A} = U$	$A \cap \bar{A} = \emptyset$	Complement law

3.36.2   Theorem 2

For all subsets, A, B, & C of a set U, the following equations are identities[19]:

6. If, for all $A, A \cup B = A$, then $B = \emptyset$	If, for all $A, A \cap B = A$, then $B = U$	Self-dual
7. $A \cup B = U$ and $A \cap B = \emptyset$, then $B = \bar{A}$		Self-dual
8. $\bar{\bar{A}} = A$		Self-dual
9. $\bar{\bar{\emptyset}} = U$	$U = \emptyset$	Identity laws
10. $A \cup A = A$	$A \cap A = A$	Idempotent law
11. $A \cup U = U$	$A \cap \emptyset = \emptyset$	Identity law
12. $A \cup (A \cap B) = A$	$A \cap (A \cup B) = A$	Absorption law
13. $\overline{A \cup B} = \bar{A} \cap \bar{B}$	$\overline{A \cap B} = \bar{A} \cup \bar{B}$	DeMorgan law

3.37   Zero Sum Marginal Difference

We will compute a *Zero Sum Marginal Difference (ZSMD)* by subtracting the *marginal* probabilities of nodes B, C, and D as computed in Table 6.2, from the computed *marginals* in the joint distributions, as seen in the tables in Chaps. 7 through 9.[20]

Acknowledgements I would like to thank Dr. Philip B. Stark, Department of Statistics, University of California, Berkeley for allowing me to use his statistical terminology and definitions liberally from his website, SticiGui, found @ http://www.stat.berkeley.edu/~stark/SticiGui/Text/gloss.htm (See Stark 2012).

---

[19]Here, $\bar{A} = U - A$.

[20]This verifies that the elements of the *joints* in the True and False events sum to zero.

References

Bolstad, W.M. (2007). *Introduction to Bayesian statistics* (2nd ed.). Hoboken, NJ: Wiley.

Disjoint sets. (n.d.). Retrieved November 06, 2016, from https://en.wikipedia.org/wiki/Disjoint_sets

Hebert, S., Lee, V., Morabito, M., & Polan, J. (2007). *Bayesian network theory*. https://controls.engin.umich.edu/wiki/index.php/Bayesian_network_theory. Accessed 23 July 2012.

Law of total probability (n.d). Retrieved November 06, 2016, from https://en.wikipedia.org/wiki/Law_of_total_probability

Stark, P. B. (2012). *Glossary of statistical terms*. http://statistics.berkeley.edu/~stark/SticiGui/Text/gloss.htm. Accessed 14 September 2016.

Statistics Glossary – probability. (n.d.). Retrieved November 06, 2016, from http://www.stats.gla.ac.uk/steps/glossary/probability.html/#event

Stoll, R. R. (1979). *Set theory and logic*. Mineola, NY: Dover Publications.

Chapter 4
Bayesian Belief Networks Experimental protocol

4.1 Introduction

This chapter presents the Bayesian statistical methodology we will use in building and invoking a series of synthetic 2, 3, and 4 node Bayesian Belief Networks (BBN). We will demonstration these steps using the concepts, and statistics, and probability theory in Chap. 3, *Statistical Properties of Bayes' Theorem.*

We seek to identify a universal set of mutually exclusive, disjoint, subsets. We see the truth of the unobservable event/s by conditioning them on the observable events, using frequencies of conditional occurrences, of invoked events from a Monte Carlo data set. When we count these frequencies, we have a starting point in probability space. We are asking, "What is the probability of an event B given the evidence of an event A, or $P(B|A)$ in a BBN?" After establishing this basic block of the theorem, we can incorporate the constructs of BBN and Bayesian updating, as illustrated in Sect. 3.5. Understanding this concept is critical, in properly interpreting the outcome probabilities.

4.2 Bayes' Research Methodology

4.2.1 Joint Space and Disjoint Bayesian Belief Network Structure

This section describes the 9-steps to use, in conducting a Bayesian experiment.[1] It begins with identifying the population of interest, and creating our sample space, which allows us to update the *priors*, using *posterior* probabilities. We do this by

---

[1]This includes joint and disjoint BBN.

© Springer International Publishing AG 2016
J. Grover, *The Manual of Strategic Economic Decision Making,*
DOI 10.1007/978-3-319-48414-3_4

Fig. 4.1 Specifying a BBN

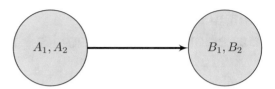

conducting a random experiment, collecting the data to create the BBN and using the *chain rule* to propagate the *posteriors*, which result in the updating of the *priors*.

> *Step 1: Identify a population of interest.* This will be a collection of studied units (Sect. 3.28).
>
> *Step 2: Specify a BBN with joint and/or disjoint Nodes. (Sect. 3.4).*[2] We slice through this population and identify a contiguous set of variables that will become the nodes for the BBN.
>
> As an example, Fig. 4.1 represents a 2-Node, joint BBN (Sect. 3.3), where A_i and B_i represent the unobservable and observable events, respectively.[3] We are looking for an observed event, where we can slice through to identify their proportional outcomes, in terms of *posterior* outcomes, which become our adjusted *priors* (Sect. 3.4).
>
> *Step 3: Slice through each node, and identify at a minimum, two mutually exclusive or disjoint (unconditional) events, which are the subsets of our population.* We evaluate the data, and select mutually exclusive events, from each node or column of data.
>
> *Step 4: Conduct the random experiment.* We will conduct a *random experiment* (See Sect. 3.30.6) and collect the data (Step 2).
>
> We will select at random one of the elements of the observable[4] variable, B_i, based on the a priori probabilities identified in Step 3. Depending on the drawn, either B or $\tilde{B}$, we will take a random sample, from the unobservable variable A_i, to select either event A or $\tilde{A}$, and assign this conditionally to obtain their initial count. We will do this iteratively until we have obtained the desired sample size.
>
> *Step 5: Determine frequency counts* We conduct joint space, conditional frequency counts of events from variables of interest, which consist of observable and unobservable, prior, events, and then we report these results in a frequency count table. Table 4.1 reports the logic, for computing these frequency counts.[5]

---

[2]We use variables (Var) and nodes interchangeably.

[3]In this manual, Var A, or Node A will always be the parent node.

[4]Since this is an unobservable event, we cannot identify the elements of this set so the conditional identification of the observable element allows for the counting.

[5]Our Var A is our proxy for the unobservable event.

Table 4.1 Joint space frequency counts

Observable events	Unobservable events		
	A	$\tilde{A}$	Marginals
B	$C(A, B)$[a]	$C(\tilde{A}, B)$	$C(B)$[b]
$\tilde{B}$	$C(A, \tilde{B})$	$C(\tilde{A}, \tilde{B})$	$C(\tilde{B})$
Marginals	$C(A)$[c]	$C(\tilde{A})$	100.0 %[d]

Note: These values are the total joint space frequency counts we obtain from the sampling process. This is a precursor, to computing the likelihood probabilities. These counts represent the results, from a random sample
[a] Represents the observable count of event A and B
[b] Represents the marginal count of $C(B) = C(A, B) + C(\tilde{A}, B)$
[c] Represents the marginal count of $C(A) = C(A, B) + C(A, \tilde{B})$
[d] Represents the total count of all observations: $C(A) + C(\tilde{A}) = 100.0\%$ and/or $C(B) + C(\tilde{B}) = 100.0\%$

Table 4.2 Prior probabilities

Unobservable events		
A	$\tilde{A}$	Total
$P(A)$[a]	$P(\tilde{A})$	100.0 %[b]

Note: These values represent the unobservable events, A and $\tilde{A}$, reported in Table 4.1
[a] $P(A) = C(A)/(C(A) + C(\tilde{A}))$
[b] $100.0\% = P(A) + P(\tilde{A})$ using the law of total probability (Sect. 3.22)

Step 6: Determine prior[6] or unconditional probabilities.[7]
These *priors* can be subjective or objective and represent the proportions that we believe exists in the sample space.[8] We calculate them by collapsing down on each element of the unobservable event, as reported in Table 4.1, and then report these results in Table 4.2.

---

[6]These *priors* can be vague, but we will see that they can be "washed them away" across BNN nodes using the chain rule of probability.

[7]In theory, we can determine these discrete event probabilities before conducting the experiment, to satisfy the independence requirement of the theorem. This is a common fallacy and researchers should not use *priors* they obtain from observable events. The concept of prior information and unconditional (see Earman (1992) for a thorough discussion of the problem of *priors* and other foundational controversies in Bayesian philosophy of science) probabilities is unique and it represents how confident we are in our initial beliefs.

[8]In our experiment, we will deviate from this premise only to provide contiguous examples of synthetic BBN and their solutions. We will obtain these *priors* by counting the experimental data.

Table 4.3 Likelihood probabilities

	Unobservable events		
Observable events	A	$\tilde{A}$	Total (%)
B	$P(A|B)$[a]	$P(\tilde{A}|B)$	100.0[b]
$\tilde{B}$	$P(A|\tilde{B})$	$P(\tilde{A}|\tilde{B})$	100.0

[a] $P(A|B) = C(A, B)/C(B)$, which represents the conditional probability of event A given B. This is computed from Table 4.1
[b] $P(B) = P(A|B) + P(\tilde{A}|B)$, which is always 100.0 %

Table 4.4 Joint and marginal probabilities

	Unobservable events		
Observable events	A	$\tilde{A}$	Marginals (%)
B	$P(B, A)$[a]	$P(B, \tilde{A})$	$P(B)$[b]
$\tilde{B}$	$P(\tilde{B}, A)$	$P(\tilde{B}, \tilde{A})$	$P(\tilde{B})$
Marginals (%)	$P(A)$[c]	$P(\tilde{A})$	100.0[d]

[a] Represents the *joint* events A and B. This is computed using the chain rule [Eq. (3.3)] as $P(A, B) = P(A|B) * P(A)$
[b] Here we are marginalizing (Sect. 3.21.2) across the unobservable event A where $P(B) = P(B, A) + P(B, \tilde{A})$
[c] The *marginals* of event A are $P(A) = P(B, A) + P(\tilde{B}, A)$
[d] $100.0\% = P(A) + P(\tilde{A}) + P(B) + P(\tilde{B})$. The total *joints* always equals 100.0 %

Step 7: Determine likelihood probabilities. We compute *likelihoods* conditioned on the conditional counts in Table 4.1. We calculate probabilities, across the unobservable event by dividing them with the total count, of each observable event. We report these results in Table 4.3.

Step 8: Determine joint and marginal probabilities. We use the chain rule (Sect. 3.8), to compute *joints*. We multiply the *priors* and *likelihoods* in Tables 4.2 and 4.3 across the observable and unobservable events. To compute *marginals*, we sum across the *joints* of the unobservable events and then down the observable events. The sum of all the *joints* always total to 100 %. We report these in Table 4.4 as *joint* and *marginals*.[9]

Step 9: Determine posterior probabilities. We compute posterior probabilities by dividing the *joints* in Table 4.4 by their *marginals*. The sum of all the *posteriors* always total to 100 %. We report these in Table 4.5 as posteriors.

---

[9]Of great importance in algorithm development, is that one can compute the joint probabilities directly from the frequency counts. For example, $P(A, B) = C(A, B)/Total$.

Table 4.5 Posterior probabilities

Unobservable events	Observable events		
	B	$\tilde{B}$	Total (%)
A	$P(B|A)$[a]	$P(\tilde{B}|A)$	100.0[b]
$\tilde{A}$	$P(B|\tilde{A})$	$P(\tilde{B}|\tilde{A})$	100.0

Note: The observable and unobservable events switch rows and columns

[a] $P(B|A) = P(A, B)/P(A)$

[b] This represents the total probability of $P(A) = P(B|A) + P(\tilde{B}|A)$, which is always 100.0 % using the total law of probability (Sect. 3.22)

Reference

Earman, J. (1992). *Bayes or bust: A critical examination of Bayesian confirmation theory.* Cambridge: MIT Press.

Chapter 5
Strategic Economic Decision-Making-I Chapter Examples

5.1 Introduction

We will draw on the concepts in Chap. 4, Bayesian Belief Networks (BBN) Experimental protocol.[1] The protocol is aimed towards the joint and disjoint BBN construction. We will re-represent the examples illustrated in Grover 2013. These examples will consist of the following:

- Manufacturing
- Political Science
- Gambling
- Publicly Traded Company Default
- Acts of Terrorism
- Currency Wars
- College Entrance Exams
- Insurance Risk Levels
- Special Forces Assessment and Selection, Two-Node Example
- Special Forces Assessment and Selection, Three-Node Example

5.1.1 Manufacturing Example

This highlights an example of a BBN in a manufacturing scenario, by evaluating the variables of "Transistor Quality" and "Suppliers." Quality control and costs have

---

This chapter is a corrected and condense version of Grover 2013.

[1] We will use subjective matter expertise subjective probabilities, for the starting values of the *prior* and the *conditional* counts for the *likelihood* probabilities.

© Springer International Publishing AG 2016
J. Grover, *The Manual of Strategic Economic Decision Making*,
DOI 10.1007/978-3-319-48414-3_5

great utility in the company's ability to make a profit, gain a competitive advantage, and maintain their reputation as an industry leader.[2]

5.1.1.1   Scenario

XYZ Electronics, Inc. obtains transistors and stores them in an open container. The assembly department is experiencing an above-average number of defective transistors, from three suppliers, Companies A, B, and C. The concern is that these defective parts will start slowing down assembly time in the plant, and exponentially increase the cost of goods sold. Their research question is to determine the proportions of defective and non-defective transistors, from these suppliers. Obtaining quality transistors, with the minimal amount of costs, will benefit XYZ Electronics, Inc.

5.1.1.2   Experimental Protocol

Step 1: Identify a population of interest. The population consists of the total number of transistors produced and supplied by Company A, B, and C, that were Non-Defective and Defective.
Step 2: Specify a BBN with joint and/or disjoint Nodes. The nodes are joint, and we have labeled them Transistor Quality and Supplier.
Step 3: Slice through each node, and identify at a minimum, two mutually exclusive or disjoint (unconditional) events, which are the subsets of our population. For the unobservable node, Transistor Quality, the disjoint events consist of Non-Defective and Defective transistors and for the observable node, Supplier, they consist of Company A, B, and C.
Step 4: Conduct the random experiment. We perform the experiment, by making random draws from Transistor Quality and then from Supplier, from separate storage containers. The sampling process starts with a single random draw from Transistor Quality, and then from Supplier, and ends with the assignment of the draw results. We will continue this process, until we have the desired sample size.
Step 5: Determine frequency counts. We will count the joint frequencies of Transistor Quality and Supplier events for further analysis. We report the results in Table 5.1 for 335 iterations.
Step 6: Determine prior or unconditional probabilities. Historically, 71.4 % of the transistors have been Non-Defective, and 28.6 % have been Defective. We report these in Table 5.2.
Step 7: Determine likelihood probabilities. We compute *likelihoods* conditioned, on the conditional counts, in Table 5.1. To determine these percentages, we calculate

---

[2]The context of this example is from Weiers et al. (2005).

Table 5.1 Total variable frequency counts

Transistor Quality	Supplier			
	Company A	Company B	Company C	Total
Non-Defective	120[a]	83	72	275
Defective	17	12	31	60
Total	137[b]	95	103	335

Note: These values represent the joint counts, for each of the events of Supplier and Transistor quality

[a]This represents the joint event counts, of Non-Defective and Company A

[b]This represents the marginal count of Company A and Non-Defective and Defective, Transistor Quality events

Table 5.2 Prior probabilities

Transistor Quality		
Non-Defective	Defective	Total
0.714000	0.286000	1.00000[a]

[a]$1.000000 = 0.286000 + 0.714000$ using the law of total probability (Sect. 3.22)

Table 5.3 Likelihood probabilities

Transistor Quality	Supplier			
	Company A	Company B	Company C	Total
Non-Defective	0.436364[a]	0.301818	0.261818	1.000000
Defective	0.283333	0.200000	0.516667	1.000000

Note: These values represent the *likelihood* for Supplier given Transistor Quality, for each event

[a]From Table 5.1, $0.436364 = 120/275$

probabilities across the observable event by dividing them with the total count of each unobservable event. We report these results in Table 5.3.

Step 8: Determine joint and marginal probabilities. To compute the *joints*, we multiply the *priors* times the *likelihoods* in Tables 5.2 and 5.3, across the events of Transistor Quality and Supplier. To compute the *marginals* of Transistor quality events, we sum across the joint events of Supplier, and to compute those of Supplier, we sum down the events of Transistor Quality. The sum of the individual *joints* totals 100.0 %. We report the results in Table 5.4.

Step 9: Determine posterior probabilities. We compute *posteriors* based on the *joints and marginals* in Table 5.4. To determine these probabilities, we divide the Supplier and Transistor Quality joint event by the respective Transistor Quality marginal event. We report the results in Table 5.5.

Table 5.4 Joint and marginal probabilities

Transistor Quality	Supplier			
	Company A	Company B	Company C	Marginal
Non-Defective	0.311564[a]	0.215498	0.186938	0.714000
Defective	0.081033	0.057200	0.147767	0.286000
Marginal	0.392597	0.272698	0.334705	1.000000

Note: These values represent the *joints* for each Supplier and Transistor Quality event, that we calculated using the *priors* and *likelihoods*
[a]From Tables 5.2 and 5.3, 0.311564 = 0.714000×0.436364

Table 5.5 Posterior probabilities

Supplier	Transistor Quality		
	Non-Defective	Defective	Total
Company A	0.793597[a]	0.206403	1.000000
Company B	0.790244	0.209756	1.000000
Company C	0.558517	0.441483	1.000000

[a]From Table 5.4, 0.793597 = 0.311564/0.392597. This represents the probability of Transistor Quality = Non-Defective given Supplier = Company A

5.1.2   Political Science Example

This highlights an example of a BBN in a national economic scenario, by evaluating the variables of "County" and "Political Affiliation." The balance between Democrats and Republicans has great utility in a politician's ability to determine and maintain her or his reputation as a government servant.

5.1.2.1   Scenario

In this scenario, an incumbent State Republican Senator obtains constituent votes from two Republican counties and hopes to maintain voter confidence. The Senator is experiencing an above-average number of party defectors from these two counties, County A and B. The Senator's concern is that these defectors will sway other voters, and jeopardize the upcoming elections, while exponentially increase the cost of winning these voters back. His or her research question is to determine the proportions of remaining political affiliations. Obtaining loyal voters with the minimal amount of cost would be a benefit to the winning strategy of the campaign.

5.1.2.2   Experimental Protocol

Step 1: Identify a population of interest. The population consists of the total number of registered voters in Counties A and B.

Step 2: Specify a BBN with joint and/or disjoint Nodes. The nodes are joint, and we have labeled them, County and Political Affiliation.

Step 3: Slice through each node, and identify at a minimum, two mutually exclusive or disjoint (unconditional) events, which are the subsets of our population. For the unobservable node, County, the disjoint events consist of County A and B, and for the observable node, Political Affiliation, they consist of Democrat and Republican.

Step 4: Conduct the random experiment. From separate databases, we perform the experiment by making random draws from County and then from Political Affiliation. The sampling process starts with a single random draw from County, and then from Political Affiliation, and ends with the assignment of draw results. We will continue this process until we have the desired sample size.

Step 5: Determine frequency counts. We will count joint the frequencies of County and Political Affiliation events for further analysis. We will report these results in Table 5.6 for 750 iterations.

Step 6: Determine prior or unconditional probabilities. Historically, 28.6 % of the constituents in County A and 71.4 % of County B have supported the incumbent. We report these in Table 5.7.

Step 7: Determine likelihood probabilities. We compute *likelihoods* conditioned, on the conditional counts, in Table 5.6. To determine these percentages, we calculate probabilities across the observable event by dividing them with the total count of each unobservable event. We report these results in Table 5.8.

Step 8: Determine joint and marginal probabilities. To compute the *joints*, we multiply the *priors* times the *likelihoods* in Tables 5.7 and 5.8 across the events of County and Political Affiliation. To compute the *marginals* of County events, we

Table 5.6 Total variable frequency counts

	Political Affiliation		
County	Democrat	Republican	Total
County A	243[a]	145	388
County B	211	151	362
Total	454[b]	296	750

Note: These values represent the joint counts, for each of the events of Political Affiliation and County

[a]This represents the joint event counts of Democrat and County A

[b]This represents the marginal count of all Democrat and County A and B, County events

Table 5.7 Prior probabilities

County		
County A	County B	Total
0.286000	0.714000	1.00000[a]

[a]1.000000 = 0.286000 + 0.714000 using the law of total probability (Sect. 3.22)

Table 5.8 Likelihood probabilities

	Political Affiliation		
County	Democrat	Republican	Total
County A	0.626289[a]	0.373711	1.000000
County B	0.582873	0.417127	1.000000

Note: These values represent the *likelihood* for Political Affiliation given County, for each event
[a]From Table 5.6, 0.626289 = 243/388

Table 5.9 Joint and marginal probabilities

	Political Affiliation		
County	Democrat	Republican	Marginal
County A	0.179119[a]	0.106881	0.286000
County B	0.416171	0.297829	0.714000
Marginal	0.595290	0.404710	1.000000

Note: These values represent the *joints* for each Political Affiliation and County event, that we calculated using the *priors* and *likelihoods*
[a]From Tables 5.7 and 5.8, 0.179119 = 0.286000×0.626289

Table 5.10 Posterior probabilities

	County		
Political Affiliation	County A	County B	Total
Democrat	0.300893[a]	0.699107	1.000000
Republican	0.264094	0.735906	1.000000

[a]From Table 5.9, 0.300893 = 0.179119/0.595290. This represents the probability of County = County A given Political Affiliation = Democrat

sum across the joint events of Political Affiliation, and to compute those of Political Affiliation, we sum down the events of County. The sum of the individual *joints* totals 100.0 %. We report the results in Table 5.9.

Step 9: Determine posterior probabilities. We compute *posteriors* based on the *joints and marginals* in Table 5.9. To determine these probabilities, we divide the Political Affiliation and County joint event by the respective County marginal event. We report the results in Table 5.10.

5.1.3   Gambling Example

This highlights an example of a BBN in a gaming scenario, by evaluating the variables of "Die Randomness" and "Fair Die." The balance between winning and losing has great utility, in a casino's ability to remain profitable, while hedging risk to maintain their reputation as a gaming establishment.

5.1.3.1 Scenario

A casino obtains die from a custom manufacturer who ensures fairness, and issues them to employees working at gaming tables. The casino experiences an above-average number of wins from its gamblers who are either using loaded or unloaded die. Their concern is that loaded die will begin eroding their profit margins. Their research question is to determine the proportions of die fairness, given the contributions of respective die. Obtaining fair die with the minimal amount of cost would be a benefit to the casino.

5.1.3.2 Experimental Protocol

Step 1: Identify a population of interest. The population consists of the total number of die throws that were Loaded and Unloaded.
Step 2: Specify a BBN with joint and/or disjoint Nodes. The nodes are joint, and we have labeled them, Die Randomness and Fair Die.
Step 3: Slice through each node, and identify at a minimum, two mutually exclusive or disjoint (unconditional) events, which are the subsets of our population. For the unobservable node, Die Randomness, the disjoint events consist of Winner and Loser, and for the observable node, Fair Die, they consist of Loaded and Unloaded.
Step 4: Conduct the random experiment. We perform the experiment by making random draws from Die Randomness and Fair Die from a database. The sampling process starts with a single random draw, from Die Randomness, then from Fair Die, and ends with the assignment of the draw results. We will continue this process until we have the desired sample size.
Step 5: Determine frequency counts. We will count joint the frequencies of Die Randomness and Fair Die events for further analysis. We report the results in Table 5.11 for 118 iterations.
Step 6: Determine prior or unconditional probabilities. Historically, the casino has experienced 50.0 % of gamblers who were Winners and 50.0 % who were Losers. We report these in Table 5.12.

Table 5.11 Total variable frequency counts

	Fair Die		
Die Randomness	Loaded	Unloaded	Total
Winner	4[a]	7	11
Loser	58	49	107
Total	62[b]	56	118

Note: These values represent the joint counts, for each of the events of Fair Die and Die Randomness
[a]This represents the joint event counts, of Loaded and Winner
[b]This represents the marginal count of all Loaded and Winner and Loser, Die Randomness events

Table 5.12 Prior
probabilities

Die Randomness		
Winners	Losers	Total
0.500000	0.500000	1.00000[a]

[a]1.000000 = 0.500000 + 0.500000
using the law of total probability
(Sect. 3.22)

Table 5.13 Likelihood
probabilities

	Fair Die		
Die Randomness	Loaded	Unloaded	Total
Winner	0.363636[a]	0.636364	1.000000
Loser	0.542056	0.457944	1.000000

Note: These values represent the *likelihood* for Fair Die
given Die Randomness, for each event
[a]From Table 5.11, 0.363636 = 4/11

Table 5.14 Joint and
marginal probabilities

	Fair Die		
Die Randomness	Loaded	Unloaded	Marginal
Winner	0.181818[a]	0.318182	0.500000
Loser	0.271028	0.228972	0.500000
Marginal	0.452846	0.547154	1.000000

Note: These values represent the *joints* for each Fair Die
and Die Randomness event, that we calculated using the
priors and *likelihood*
[a]From Tables 5.12 and 5.13, 0.181818 = 0.500000 ×
0.363636

Step 7: Determine likelihood probabilities. We compute *likelihoods* conditioned on
the conditional counts, in Table 5.11. To determine these percentages, we calculate
probabilities, across the observable event, by dividing them with the total count of
each unobservable event, to report these results in Table 5.13.

Step 8: Determine joint and marginal probabilities. To compute the *joints*, we
multiply the *priors* times the *likelihoods* in Tables 5.12 and 5.13 across the events
of Die Randomness and Fair Die. To compute the *marginals* of Die Randomness
events, we sum across the joint events of Fair Die, and to compute those of Fair Die,
we sum down the events of Die Randomness. The sum of the individual *joints* totals
100.0 %. We report the results in Table 5.14.

Step 9: Determine posterior probabilities. We compute *posteriors* based on the
joints and marginals in Table 5.14. To determine these probabilities, we divide the
Fair Die and Die Randomness joint event by the Die Randomness marginal event.
We report the results in Table 5.15.

Table 5.15 Posterior
probabilities

Fair Die	Die Randomness		
	Winner	Loser	Total
Loaded	0.401501[a]	0.598499	1.000000
Unloaded	0.581522	0.418478	1.000000

[a]From Table 5.14, 0.401501 = 0.181818/0.452846. This represents the probability of Die Randomness = Winner given Fair Die = Loaded

5.1.4 Publicly Traded Company Example

This highlights an example of a BBN in a national economic scenario, by evaluating the variables of "Altman Z-Scores" and "Health Status." The balance between international company default, and investments has great economic utility in a country's ability to warn its international investors, and to hedge global effects of default to maintain their reputation as a global financial leader.

5.1.4.1 Scenario

The United States (U.S.) Office of the Controller (OCC) has issued a warning that the US Dollar (USD) may be devalued. The OCC maintains facility ratings of multinational corporations (MNC) operation in the U.S. and use the Altman Z-Score to evaluate their probability of remaining as an ongoing concern. Several of these corporations have low ratings, which is a growing trend. The New Your Stock Exchange (NYSE) have them listed and they greatly influence the USD. The research question is this: What are the proportions of Z-Scores assigned to these MNC given they remain as going concerns, transition in merger or acquisition (M&A), are dissolved, or go bankrupt.

5.1.4.2 Experimental Protocol

Step 1: Identify a population of interest. The population consists of the total number of current and historical, publicly traded companies listed on the NYSE with Z-Score ≤ 3 & Z-Score > 3.

Step 2: Specify a BBN with joint and/or disjoint Nodes. The nodes are joint, and we have labeled them, Z-Score and Health Status.

Step 3: Slice through each node, and identify at a minimum, two mutually exclusive or disjoint (unconditional) events, which are the subsets of our population. For the unobservable node, Z-Score, the disjoint events consist of Z-Score ≤ 3 and Z-Score > 3, and for the observable node, Health Status, they consist of Going Concern, M&A, Dissolved, and Bankrupt.

Table 5.16 Total variable frequency counts

Z-Score	Health Status				
	Going Concern	M&A	Dissolved	Bankrupt	Total
Z-Score ≤ 3	256[a]	583	1978	239	3056
Z-Score > 3	6439	459	321	152	7371
Total	6695[b]	1042	2299	391	10,427

Note: These values represent the joint counts, for each of the events of Health Status and Z-Score
[a]This represents the joint event counts, of Going Concern and Z-Score ≤ 3
[b]This represents the marginal count of all Going Concern and Z-Score ≤ 3 and Z-Score > 3, Z-Score events

Table 5.17 Prior probabilities

Z-Score		
Z-Score ≤ 3	Z-Score > 3	Total
0.050000	0.950000	1.00000[a]

[a]$1.000000 = 0.050000 + 0.950000$ using the law of total probability (Sect. 3.22)

Step 4: Conduct the random experiment. We perform the experiment by making random draws from Z-Score and then from Health Status, from a database of companies that have been listed on the NYSE. The sampling process starts with a single random draw from Z-Score, and then from Health Status, and ends with the assignment of the draw results. We will continue this process until we have the desired sample size.

Step 5: Determine frequency counts. We will count joint the frequencies of Z-Score and Health Status events for further analysis. We report the results in Table 5.16 for 10,427 iterations.

Step 6: Determine prior or unconditional probabilities. Historically, the NYSE has experienced 5.0 % of the listed publicly traded companies with Z-Score ≤ 3 and 95.0 % with Z-Score > 3. We report these in Table 5.17.

Step 7: Determine likelihood probabilities. We compute *likelihoods* conditioned, on the conditional counts, in Table 5.16. To determine these percentages, we calculate probabilities, across the observable event by dividing them with the total count of each unobservable event. We report these results in Table 5.18.

Step 8: Determine joint and marginal probabilities. To compute the *joints*, we multiply the *priors* times the *likelihoods* in Tables 5.17 and 5.18 across the events of Z-Score and Health Status. To compute the *marginals* of Z-Score events, we sum across the joint events of Health Status. To compute those of Health Status, we sum down the events of Z-Score. The sum of the individual *joints* totals 100.0 %. We report the results in Table 5.19.

Table 5.18 Likelihood probabilities

	Health Status				
Z-Score	Going Concern	M&A	Dissolved	Bankrupt	Total
Z-Score ≤ 3	0.083770[a]	0.190772	0.647251	0.078207	1.000000
Z-Score > 3	0.873559	0.062271	0.043549	0.020621	1.000000

Note: These values represent the *likelihood* for Health Status given Z-Score, for each event
[a]From Table 5.16, 0.083770 = 256/3056

Table 5.19 Joint and marginal probabilities

	Health Status				
Z-Score	Going Concern	M&A	Dissolved	Bankrupt	Marginal
Z-Score ≤ 3	0.004188[a]	0.009539	0.032363	0.003910	0.050000
Z-Score > 3	0.829881	0.059158	0.041372	0.019590	0.950000
Marginal	0.834069	0.068696	0.073734	0.023501	1.000000

Note: These values represent the *joints* for each Health Status and Z-Score event, that we calculated using the *priors* and *likelihoods*
[a]From Tables 5.17 and 5.18, where 0.004188 = 0.05000×0.083770

Table 5.20 Posterior probabilities

	Z-Score		
Health Status	Z-Score ≤ 3	Z-Score > 3	Total
Going Concern	0.005022[a]	0.994978	1.000000
M&A	0.138852	0.861148	1.000000
Dissolved	0.438909	0.561091	1.000000
Bankrupt	0.166393	0.833607	1.000000

[a]From Table 5.19, 0.005022 = 0.004188/0.834069. This represents the probability of Z-Score = Z-Score ≤ 3 given Health Status = Going Concern

Step 9: Determine posterior probabilities. We compute *posteriors* based on the *joints and marginals* in Table 5.19. To determine these probabilities, we divide the Health Status and Z-Score joint event, by the respective Z-Score, marginal event. We report the results in Table 5.20.

5.1.5   Insurance Example

This highlights an example of a BBN in an insurance scenario by evaluating the variables of "Risk Category" and "Fatality Status." The balance between risk and premiums have great economic utility in the company's ability to make a profit, gain a competitive advantage, and maintain their reputation as an industry leader.

5.1.5.1 Scenario

An insurance company obtains insurer risk categories for multiple fatality statuses and provides insurers with respective rate quotes. The agency is experiencing an above-average number of discounted quotes, that do not hedge the risk of respective insurers. Their concern is these discounted quotes will start putting pressure on general policy-holders to hedge forecasted pay-outs to fatality insurers and increasing the cost of policies sold exponentially. Their research question is to determine the proportions of age group insurance premiums given the results of future fatality statuses. Obtaining optimal insurance pricing with the minimal amount of costs, would be a benefit to the insurance company and its agents.

5.1.5.2 Experimental Protocol

Step 1: Identify a population of interest. The population consists of the total number of risk categories that included Age groups 16–19, 20–24, and > 25.

Step 2: Specify a BBN with joint and/or disjoint Nodes. The nodes are joint, and we have labeled them, Risk Category and Fatality Status.

Step 3: Slice through each node, and identify at a minimum, two mutually exclusive or disjoint (unconditional) events, which are the subsets of our population. For the unobservable node, Risk Category, the disjoint events consist of age brackets of Age 16–19, Age 20–24, and Age $\geq$ 25, and for the observable node, Fatality Status, they consist of Fatality and No fatality.

Step 4: Conduct the random experiment. We perform the experiment by making random draws from Risk Category and then from Fatality Status from an insurer database. The sampling process starts with a single random draw from Risk Category, then from Fatality Status, and ends with the assignment of the draw results. We will continue this process until we have the desired sample size.

Step 5: Determine frequency counts. We will count joint the frequencies of Risk Category and Fatality Status events for further analysis. We report the results in Table 5.21 for 2649 iterations.

Step 6: Determine prior or unconditional probabilities. Historically, the insurance company has reported insurer Risk Category rates of 45.0 % for Age Group 16–19, 35.0 % for Age Group 20–24, and 20.0 % for Age Group $\geq$ 25. We report these in Table 5.22.

Step 7: Determine likelihood probabilities. We compute *likelihoods* conditioned, on the conditional counts in Table 5.21. To determine these percentages, we calculate probabilities across the observable event by dividing them with the total count of each unobservable event. We report these results in Table 5.23.

Table 5.21 Total variable frequency counts

	Fatality Status		
Risk Category	Fatality	No Fatality	Total
Age 16–19	269[a]	838	1107
Age 20–24	273	649	922
Age ≥ 25	59	561	620
Total	601[b]	2048	2649

Note: These values represent the joint counts, for each of the events of Fatality Status and Risk Category

[a]This represents the joint event counts, of Fatality and Age 16–19

[b]This represents the marginal count of all Fatality and Age 16–19, Age 20–24, and Age ≥ 25, Risk Category events

Table 5.22 Prior probabilities

Risk Category			
Age 16–19	Age 20–24	Age ≥ 25	Total
0.450000	0.350000	0.200000	1.00000[a]

[a]$1.000000 = 0.450000 + 0.350000 + 0.200000$ using the law of total probability (Sect. 3.22)

Table 5.23 Likelihood probabilities

	Fatality Status		
Risk Category	Fatality	No Fatality	Total
Age 16–19	0.242999[a]	0.757001	1.000000
Age 20–24	0.296095	0.703905	1.000000
Age ≥ 25	0.095161	0.904839	1.000000

Note: These values represent the *likelihood* for Fatality Status given Risk Category, for each event

[a]From Table 5.21, $0.242999 = 269/1107$

Step 8: Determine joint and marginal probabilities. To compute the *joints*, we multiply the *priors* times the *likelihoods* in Tables 5.22 and 5.23 across the events of Risk Category and Fatality Status. To compute the *marginals* of Risk Category events, we sum across the joint events of Fatality Status, and to compute those of Fatality Status, we sum down the events of Risk Category. The sum of the individual *joints* totals 100.0 %. We report the results in Table 5.24.

Step 9: Determine posterior probabilities. We compute *posteriors* based on the *joints and marginals* in Table 5.24. To determine these probabilities, we divide the Fatality Status and Risk Category joint event by the respective Risk Category marginal event. We report the results in Table 5.25.

Table 5.24 Joint and marginal probabilities

	Fatality Status		
Risk Category	Fatality	No Fatality	Marginals
Age 16–19	0.109350[a]	0.340650	0.450000
Age 20–24	0.103633	0.246367	0.350000
Age ≥ 25	0.019032	0.180968	0.200000
Marginals	0.232015	0.767985	1.000000

Note: These values represent the *joints* for each Fatality Status and Risk Category event, that we calculated using the *priors* and *likelihoods*
[a]From Tables 5.22 and 5.23, 0.109350 = 0.450000×0.242999

Table 5.25 Posterior probabilities

	Risk Category			
Fatality Status	Age 16–19	Age 20–24	Age ≥ 25	Total
Fatality	0.471303[a]	0.446666	0.082030	1.000000
No Fatality	0.443564	0.320796	0.235640	1.000000

[a]From Table 5.24, 0.471303 = 0.109350/0.232015. This represents the probability of Risk Category = Age 16–19 given Fatality Status = Fatality

5.1.6   Acts of Terrorism Example

This highlights an example of a BBN in a hostile scenario, by evaluating the variables of "Country" and "Fatality Status." Citizen safety has great economic utility in a person's ability to maintain safety while living and traveling abroad in countries with terrorist cells whose desire is to launch their economic and political agenda on innocent citizens.

5.1.6.1   Scenario

There continues to be a threat abroad on U.S citizens by terrorist organizations. Attacks appear to be random and are on the increase globally; especially in the European countries of France, Germany, and Greece. The Office of the Secretary of State (DOS) monitors each occurrence of Acts of Terrorism (AOT) against U.S. citizens. The DOS' concerns are providing an early warning alert for U.S. citizens traveling to high-risk countries. Their research question is to determine the proportions of attacks on its citizens across these countries of concern given their severity. Having a terrorism warning tool to prevent harm or death is a benefit to the U.S. and its citizens.

5.1.6.2 Experimental Protocol

Step 1: Identify a population of interest. The population consists of the total number of countries that have had AOT resulting in fatalities, injuries, and no harm.

Step 2: Specify a BBN with joint and/or disjoint Nodes. The nodes are joint, and we have labeled them, Country and Fatality Status.

Step 3: Slice through each node, and identify at a minimum, two mutually exclusive or disjoint (unconditional) events, which are the subsets of our population. For the unobservable node, Country, the disjoint events consist of France, Germany, and Greece, and for the observable node, Fatality Status, they consist of Fatality, Injured, and No harm.

Step 4: Conduct the random experiment. We perform the experiment, by making random draws from Country, and then from Fatality Status, from an AOT database. The sampling process starts with a single random draw from Country and then from Fatality Status and ends with the assignment of the draw results. We will continue this process, until we have the desired sample size.

Step 5: Determine frequency counts. We will count joint the frequencies of Country and Fatality Status for further analysis. We report the results in Table 5.26 for 1982 iterations.

Step 6: Determine prior or unconditional probabilities. Historically, DOS has reported embassy AOT of 21.0 % in France, 59.0 % in Germany, and 20.0 % in Greece. We report these in Table 5.27.

Step 7: Determine likelihood probabilities. We compute *likelihoods* conditioned, on the conditional counts, in Table 5.26. To determine these percentages, we calculate probabilities, across the observable event, by dividing them, with the total count, of each unobservable event, to report these results in Table 5.28.

Step 8: Determine joint and marginal probabilities. To compute the *joints*, we multiply the *priors* times the *likelihoods* in Tables 5.27 and 5.28, across the events of Country and Fatality Status. To compute the *marginals* of Country events, we sum across the joint events of Fatality Status, and to compute those of Fatality Status, we

Table 5.26 Total variable frequency counts

| Country | Fatality Status | | | |
	Fatality	Injured	No harm	Total
France	34[a]	68	590	692
Germany	12	23	650	685
Greece	21	89	495	605
Total	67[b]	180	1735	1982

Note: These values represent the joint counts, for each of the events of the Fatality Status and Country
[a]This represents the joint event counts, of Fatality and France
[b]This represents the marginal count of all Fatality and France, Germany, and Greece, Country events

Table 5.27 Prior
probabilities

Country			
France	Germany	Greece	Total
0.210000	0.590000	0.200000	1.00000[a]

[a]1.000000 = 0.210000 + 0.590000 + 0.200000
using the law of total probability (Sect. 3.22)

Table 5.28 Likelihood
probabilities

Country	Fatality Status			
	Fatality	Injured	No harm	Total
France	0.049133[a]	0.098266	0.852601	1.000000
Germany	0.017518	0.033577	0.948905	1.000000
Greece	0.034711	0.147107	0.818182	1.000000

Note: These values represent the *likelihood* for Fatality Status
given Country, for each event
[a]From Table 5.26, 0.049133 = 34/692

Table 5.29 Joint and
marginal probabilities

Country	Fatality Status			
	Fatality	Injured	No harm	Marginal
France	0.010318[a]	0.020636	0.179046	0.210000
Germany	0.010336	0.019810	0.559854	0.590000
Greece	0.006942	0.029421	0.163636	0.200000
Marginal	0.027596	0.069868	0.902537	1.000000

Note: These values represent the *joints* for each Fatality
Status and Country event, that we calculated using the *priors*
and *likelihoods*
[a]From   Tables   5.27   and   5.28,   0.010318   =
0.210000×0.049133

sum down the events of Country. The sum of the individual *joints* totals 100.0 %.
We report the results in Table 5.29.

Step 9: Determine posterior probabilities. We compute *posteriors* based on the
joints and marginals in Table 5.29. To determine these probabilities, we divide the
Fatality Status and Country joint event by the respective Country marginal event.
We report the results in Table 5.30.

5.1.7   Currency Wars Example

This highlights an example of a BBN in an investment scenario by evaluating
the variables of "Currency Pair" and "Economic Effects." The economic value of
currency has great economic utility for a country's ability to provide the correct

Table 5.30 Posterior probabilities

| Fatality Status | Country | | | |
	France	Germany	Greece	Total
Fatality	0.373894[a]	0.374541	0.251565	1.000000
Injured	0.295357	0.283540	0.421104	1.000000
No harm	0.198381	0.620312	0.181307	1.000000

[a]From Table 5.29, 0.373894 = 0.010318/0.027596. This represents the probability of Country = France given Fatality Status = Fatality

economic balance of goods and services to maintain their reputation as a solvent nation within the global free-market economy.[3]

5.1.7.1   Scenario

The U.S. dollar (USD) and currencies that are pegged to the USD continually appreciate and depreciate based on global economic conditions. The USD pegged currencies are experiencing depreciation across a range of economic effects, due to whipsaw actions that generate 1000 pip devaluation movements over short periods of time. The Office of the Controller of the Currency's (OCC) concern is that these currencies will continue to depreciate exponentially in a global market place. Their research will determine the proportions of country currency depreciation rates, given future economic effects. Maintaining optimal currency valuations with the minimal amount of economic costs would be a benefit to the countries of concern.

5.1.7.2   Experimental Protocol

Step 1: Identify a population of interest. The population consists of the total number of currency pairs of USD/CAD, EUR/USD, and USD/JPY.
Step 2: Specify a BBN with joint and/or disjoint Nodes. The nodes are joint, and we have labeled them, Currency Pair and Economic Effect.
Step 3: Slice through each node, and identify at a minimum, two mutually exclusive or disjoint (unconditional) events, which are the subsets of our population. For the unobservable node, Currency Pair, the disjoint events consist of the currency pairs of USD/CAD, EUR/USD, and USD/JPY, and for the observable node, Economic Effect, they consist of Natural disaster, Assassination, and NBI.

---

[3]The idea of a currency war comes from Rickard (2012). We obtained the data from the OANDA website (OANDA, 2016).

Table 5.31 Total variable frequency counts

	Economic Effect			
Currency Pair	Natural Disaster	Assassination	NBI	Marginals
USD/CAD	4[a]	3	8	15
EUR/USD	2	5	8	15
USD/JPY	8	2	7	17
Marginals	14[b]	10	23	47

Note: These values represent the joint counts, for each of the events of
Economic Effect and Currency Pair

[a]This represents the joint event counts, of Natural Disaster and
USD/CAD

[b]This represents the marginal count of all Natural Disaster and
USD/CAD, EUR/USD, and USD/JPY, Currency Pair events

Table 5.32 Prior
probabilities

Currency Pair			
USD/CAD	EUR/USD	USD/JPY	Marginals
0.220000	0.360000	0.420000	1.00000[a]

[a]$1.000000 = 0.220000 + 0.360000 + 0.420000$
using the law of total probability (Sect. 3.22)

Step 4: Conduct the random experiment. We perform the experiment, by making
random draws of events from Currency Pair, and then from Economic Effect, from
the Oanda database.[4] The sampling process starts with a single random draw from
Currency Pair, then from Economic Effect, and ends with the assignment of the
draw results. We will continue this process, until we have obtained the sample size.

Step 5: Determine frequency counts. We will count joint the frequencies of Currency
Pair and Economic Effect events for further analysis. We report the results in
Table 5.31 for 47 iterations.

Step 6: Determine prior or unconditional probabilities. Historically, this basket of
currencies has experienced the following pip movement mix: USD/CAD, 22.0 %,
EUR/USD, 36.0 %, and USD/JPY, 42.0 %. We report these in Table 5.32.

Step 7: Determine likelihood probabilities. We compute *likelihoods* conditioned, on
the conditional counts, in Table 5.31. To determine these percentages, we calculate
probabilities across the observable event by dividing them with the total count of
each unobservable event. We report these results in Table 5.33.

---

[4]We obtained currency data from http://www.oanda.com/.

Table 5.33 Likelihood probabilities

| Currency Pair | Economic Effect | | | |
	Natural Disaster	Assassination	NBI	Total
USD/CAD	0.266667[a]	0.200000	0.533333	1.000000
EUR/USD	0.133333	0.333333	0.533333	1.000000
USD/JPY	0.470588	0.117647	0.411765	1.000000

Note: These values represent the *likelihood* for Economic Effect given Currency Pair, for each event
[a]From Table 5.31, $0.266667 = 4/15$

Table 5.34 Joint and marginal probabilities

| Currency Pair | Economic Effect | | | |
	Natural Disaster	Assassination	NBI	Total
USD/CAD	0.058667[a]	0.044000	0.117333	0.220000
EUR/USD	0.048000	0.120000	0.192000	0.360000
USD/JPY	0.197647	0.049412	0.172941	0.420000
Marginals	0.304314	0.213412	0.482275	1.000000

Note: These values represent the *joints* for each Economic Effect and Currency Pair event, that we calculated using the *priors* and *likelihood*
[a]From Tables 5.32 and 5.33, $0.058667 = 0.220000 \times 0.266667$

Table 5.35 Posterior probabilities

| Economic Effect | Currency Pair | | | |
	USD/CAD	EUR/USD	USD/JPY	Total
Natural Disaster	0.192784[a]	0.157732	0.649485	1.000000
Assassination	0.206174	0.562293	0.231533	1.000000
NBI	0.243292	0.398114	0.358595	1.000000

[a]From Table 5.34, $0.192784 = 0.058667/0.304314$. This represents the probability of Currency Pair USD/CAD given Economic Effect = Natural Disaster

Step 8: Determine joint and marginal probabilities. To compute the *joints*, we multiply the *priors* times the *likelihoods* in Tables 5.32 and 5.33, across the events of Currency Pair and Economic Effect. To compute the *marginals* of Currency Pair events, we sum across the joint events of Economic Effect, and to compute those of Economic Effect, we sum down the events of Currency Pair. The sum of the individual *joints* totals 100.0 %. We report the results in Table 5.34.

Step 9: Determine posterior probabilities. We compute *posteriors* based on the *joints and marginals* in Table 5.34. To determine these probabilities, we divide the Currency Pair and Country joint event by the respective Country marginal event. We report the results in Table 5.35.

5.1.8   College Entrance Exams Example

This highlights an example of a BBN in an academic scenario by evaluating the variables of "Freshman Status" and "ACT Scores." Student retention at the university is paramount profit, accreditation, high academic standards, and institution reputation.

5.1.8.1   Scenario

A university obtains freshmen students based on American College Testing (ACT) score levels. The admissions department is experiencing an above-average number of freshmen dropouts, across each level of ACT scores. Their concern is that these freshmen will continue to drop out, and increase the opportunity lost cost of these students and also send a signal to the accreditation authority of possible creditability issues. Their research question is to determine the proportions of freshman maturation given the levels of ACT scores. Obtaining quality freshman with the minimal amount of cost would be a benefit to the university.

5.1.8.2   Experimental Protocol

Step 1: Identify a population of interest. The population consists of the total number of freshmen who are Graduate and Undergraduate.

Step 2: Specify a BBN with joint and/or disjoint Nodes. The nodes are joint, and we have labeled them, Freshman Status and ACT Score.

Step 3: Slice through each node, and identify at a minimum, two mutually exclusive or disjoint (unconditional) events, which are the subsets of our population. For the unobservable node, Freshman Status, the disjoint events consist of Graduate and Non-graduate, and for the observable node, ACT Score, they consist of Level 1, Level 2, and Level 3.

Step 4: Conduct the random experiment. We perform the experiment by making random draws from Freshman Status and then from ACT Score from a university database. The sampling process starts with a single random draw from Freshman Status, then from ACT Score, and ends with the assignment of the draw results. We will continue this process until we have the desired sample size.

Step 5: Determine frequency counts. We will count joint the frequencies of Freshman Status and ACT Score events for further analysis. We report the results in Table 5.36 for 626 iterations.

Step 6: Determine prior or unconditional probabilities. Historically, 85.0 % of Freshman who have been accepted were Graduates and 15.0 % were Non-Graduates. We report these in Table 5.37.

Step 7: Determine likelihood probabilities. We compute *likelihoods* on the conditional counts in Table 5.36. To determine these percentages, we calculate

Table 5.36 Total variable frequency counts

	ACT Score			
Freshman Status	Level 1	Level 2	Level 3	Total
Graduate	369[a]	52	38	459
Non-graduate	83	58	26	167
Total	452[b]	110	64	626

Note: These values represent the joint counts, for each of the events of ACT Score and Freshman Status
[a]This represents the joint event counts, of Level 1 and Graduate
[b]This represents the marginal count of all Level 1 and Graduate and Non-graduate, Freshman Status events

Table 5.37 Prior probabilities

Freshman Status		
Graduate	Non-graduate	Total
0.850000	0.150000	1.00000[a]

[a]$1.000000 = 0.850000 + 0.150000$ using the law of total probability (Sect. 3.22)

Table 5.38 Likelihood probabilities

	ACT Score			
Freshman Status	Level 1	Level 2	Level 3	Total
Graduate	0.803922	0.113290	0.082789	1.000000
Non-graduate	0.497006[a]	0.347305	0.155689	1.000000

Note: These values represent the *likelihood* for ACT Score given Freshman Status, for each event
[a]From Table 5.36, $0.803922 = 369/459$

probabilities, across the observable event, by dividing them, with the total count, of each unobservable event, to report these results in Table 5.38.

Step 8: Determine joint and marginal probabilities. To compute the *joints*, we multiply the *priors* times the *likelihoods* in Tables 5.37 and 5.38 across the events of Freshman Status and ACT Score. To compute the *marginals* of Freshman Status events, we sum across the joint events of ACT Score, and to compute those of ACT Score, we sum down the events of Freshman Status. The sum of the individual *joints* totals 100.0 %. We report the results in Table 5.39.

Step 9: Determine posterior probabilities. We compute *posteriors*, based on the *joints and marginals* in Table 5.39. To determine these probabilities, we divide the Freshman Status and ACT Score joint event by the respective ACT Score marginal event. We report the results in Table 5.40.

Table 5.39 Joint and marginal probabilities

Freshman Status	ACT Score			
	Level 1	Level 2	Level 3	Marginal
Graduate	0.683333[a]	0.096296	0.070370	0.850000
Non-graduate	0.074551	0.052096	0.023353	0.150000
Marginal	0.757884	0.148392	0.093724	1.000000

Note: These values represent the *joints* for each ACT Scores and Freshman Status event, that we calculated using the *priors* and *likelihood*

[a]From Tables 5.37 and 5.38, $0.683333 = 0.850000 \times 0.803922$

Table 5.40 Posterior probabilities

ACT Score	Freshman Status		
	Graduate	Non-graduate	Total
Level 1	0.901633[a]	0.098367	1.000000
Level 2	0.648931	0.351069	1.000000
Level 3	0.750828	0.249172	1.000000

[a]From Table 5.39, $0.901633 = 0.683333/0.757884$. This represents the probability of ACT Scores = Graduate given Freshman Status = Level 1

5.1.9   Special Forces Assessment and Selection Two-Node Example

This highlights an example of a BBN in a military scenario by evaluating the variables of "Graduate" and "Status." The Special Forces Assessment and Selection (SFAS) has great human capital economic utility in providing the U.S. Army and the special operations communities with the finest candidates available from a limited pool of Soldiers.[5,6]

5.1.9.1   Scenario

The U.S. Army Special Forces Command's (USASFC) SFAS course obtains Soldiers from the Army. The SFAS has experienced an elevated level of attrition rates

---

[5]Special Forces (2016).

[6]The SFAS course is a 3-week evaluation of enlisted and Officer Soldiers' physical, mental, and psychological capabilities to determine if they would fit the ranks of special operations Soldiers. Those Soldiers accepted through SFAS will attend either the Officer or Enlisted Special Forces Qualification Course (SFQC), for final selection, to earn the Green Beret. SFAS is only a gateway to the SFQC.

of Soldiers they are recruiting from the enlisted and officer ranks. Their concern is that the high attrition rates will stop the Special Forces community from being fully mission capable, according to regulatory requirements, and increase the cost of recruiting and assessing future Soldiers. Their research question will determine the proportions of Selected or Non-Selected enlisted and/or officer Soldiers. Selecting the right Soldier with the minimal amount of costs would benefit the U.S. Army's recruiting program.[7,8]

5.1.9.2 Experimental Protocol

Step 1: Identify a population of interest. The population consists of the total number of graduates were Selected and Non-Selected.

Step 2: Specify a BBN with joint and/or disjoint Nodes. The nodes are joint, and we have labeled them, Graduate and Status.

Step 3: Slice through each node, and identify at a minimum, two mutually exclusive or disjoint (unconditional) events, which are the subsets of our population. For the unobservable node, Graduate, the disjoint events consist of Selected and Non-selected, and for the observable node, Status they consist of Enlisted and Officer.

Step 4: Conduct the random experiment. We perform the experiment by making random draws from Graduate and Status, from a U.S. Army Soldier personnel database. The sampling process starts with a single random draw from Graduate, and from Status, and ends with the assignment of the draw results. We will continue this process until we have the desired sample size.

Step 5: Determine frequency counts. We will count joint the frequencies of Graduate and Status events for further analysis. We report the results in Table 5.41, for 1000 iterations.

Step 6: Determine prior or unconditional probabilities. Historically, 30.0 % of all Soldiers who are accepted and enter into SFAS are Selected and 70.0 % are Non-selected. We report these in Table 5.42.

Step 7: Determine likelihood probabilities. We compute *likelihoods* conditioned, on the conditional counts, in Table 5.41. To determine these percentages, we calculate probabilities across the observable event by dividing them with the total count of each unobservable event. We report these results in Table 5.43.

---

[7]See the U.S. Army's Special Forces website: http://www.sorbrecruiting.com/ (Special Forces , 2016). Last accessed: 5/17/2016.

[8]The SFAS course is a 3-week evaluation of enlisted and Officer Soldiers' physical, mental, and psychological capabilities to determine if they would fit the ranks of special operations Soldiers. Those Soldiers accepted through SFAS will attend either the Officer or Enlisted Special Forces Qualification Course (SFQC), for final selection, to earn the Green Beret. SFAS is only a gateway to the SFQC.

Table 5.41 Total variable
frequency counts

	Status		
Graduate	Enlisted	Officer	Total
Selected	663[a]	52	715
Non-selected	263	22	285
Total	926[b]	74	1000

Note: These values represent the joint counts, for each of the events of Status and Graduate
[a]This represents the joint event counts, of Enlisted and Selected
[b]This represents the marginal count of all Enlisted and Selected and Non-selected, Graduate events

Table 5.42 Prior
probabilities

Graduate		
Selected	Non-selected	Total
0.30000	0.70000	1.00000[a]

[a]1.00000 = 0.30000 + 0.70000 using the law of total probability (Sect. 3.22)

Table 5.43 Likelihood
probabilities

	Status		
Graduate	Enlisted	Officer	Total
Selected	0.922807[a]	0.077193	1.000000
Non-selected	0.927273	0.072727	1.000000

Note: These values represent the *likelihood* for Status given Graduate, for each event
[a]From Table 5.41, 0.922807 = 663/715

Step 8: Determine joint and marginal probabilities. To compute the *joints*, we multiply the *priors* times the *likelihoods* in Tables 5.42 and 5.43 across the events of Graduate and Status. To compute the *marginals* of Graduate events, we sum across the joint events of Status, and to compute those of Status, we sum down the events of Graduate. The sum of the individual *joints* totals 100.0 %. We report the results in Table 5.44.

Step 9: Determine posterior probabilities. We compute *posteriors* based on the *joints and marginals* in Table 5.44. To determine these probabilities, we divide the Graduate and Status joint event by the respective Status marginal event. We report the results in Table 5.45.

5.1.10   *Special Forces Assessment and Selection Three-Node Example*

This is an extension of the previous Two-Node example and is a Three-Node example of a BBN, using the same scenario, by evaluating the variables of

Table 5.44 Joint and
marginal probabilities

	Status		
Graduate	Enlisted	Officer	Marginal
Selected	0.278182[a]	0.021818	0.300000
Non-selected	0.645965	0.054035	0.700000
Marginal	0.924147	0.075853	1.000000

Note: These values represent the *joints* for each
Status and Graduate event that we calculated using
the *priors* and *likelihoods*
[a]From Tables 5.42 and 5.43, 0.278182 =
0.30000×0.922807

Table 5.45 Posterior
probabilities

	Graduate		
Status	Selected	Non-selected	Total
Enlisted	0.301015[a]	0.698985	1.000000
Officer	0.287637	0.712363	1.000000

[a]From Table 5.44, 0.301015 = 0.278182/
0.924147. This represents the probability of Graduate = Selected given Status = Enlisted

"Graduate," "Status," and "PT" (Physical Fitness Levels). Again, the SFAS has
great human capital economic utility in providing the U.S. Army and the special
operations communities with the finest candidates available.

5.1.10.1 Scenario

The USASFC SFAS course obtains Soldiers from the Army. The SFAS has
experienced an elevated level of attrition rates of Soldiers they are recruiting from
the enlisted and officer ranks. Their concern is that the high attrition rates will
stop the Special Forces community from being fully mission capable, according
to regulatory requirements, and increase the cost of recruiting and assessing future
Soldiers. Their research question will determine the proportions of Selected or Non-
Selected enlisted and/or officer Soldiers.

5.1.10.2 Experimental Protocol

Step 1: Identify a population of interest. The population consists of the total number
of graduates who were Selected or Non-Selected.
Step 2: Specify a BBN with joint and/or disjoint Nodes. The nodes are joint and we
have labeled them Status, Graduate, and PT.
*Step 3: Slice through each node, and identify at a minimum, two mutually exclusive
or disjoint (unconditional) events, which are the subsets of our population.* For
the unobservable node, Graduate, the disjoint events consist of Selected and

Table 5.46 Total variable frequency counts

Graduate	Status		
	Enlisted	Officer	Total
Selected	263[a]	22	285
Non-selected	663	52	715
Total	926[b]	74	1000

Note: These values represent the joint counts, for each of the events of Status and Graduate

[a]This represents the joint event counts, of Enlisted and Selected

[b]This represents the marginal count of all Enlisted and Selected and Non-selected, Graduate events

Table 5.47 Total variable frequency counts

Status	Graduate	PT			
		Above	Extreme	Average	Total
Enlisted	Selected	92[a]	82	89	263
Enlisted	Non-selected	237	234	192	663
Officer	Selected	9	6	7	22
Officer	Non-selected	18	16	18	52
Total		356[b]	338	306	1000

Note: These values represent the joint counts, for each of the events of PT, Status and Graduate

[a]This represents the joint event counts, of Above, and Enlisted and Selected

[b]This represents the joint event counts, of Above, and Enlisted and Selected events

Non-selected and for the observable node, Status, they consist of Enlisted and Officer, and for the other observable node, PT, they consist of Above, Extreme, and Average.

Step 4: Conduct the random experiment. We perform the experiment, by making random draws from Graduate and Status from a U.S. Army Soldier personnel database. The sampling process starts with a single random draw from Graduate, then from Status, and then from PT, and ends with the assignment of the draw results. We will continue this process until we have the desired sample size.

Step 5a: Determine frequency counts. We will count joint the frequencies of Graduate and Status events for further analysis. We report the results in Table 5.46 for 1000 iterations.

Step 5b: Determine frequency counts. We will count joint the frequencies of Status, Graduate, and PT events for further analysis. We report the results in Table 5.47 for 1000 iterations.

Step 6: Determine prior or unconditional probabilities. Historically, 28.5 % for all Soldiers who are accepted and enter into SFAS are Selected and 71.5 % are Non-selected. We report these in Table 5.48.

Table 5.48 Prior probabilities

Graduate		
Selected	Non-selected	Total
0.285000	0.715000	1.00000[a]

[a] 1.00000 $=$ 0.285000 $+$ 0.715000 using the law of total probability (Sect. 3.22)

Table 5.49 Likelihood probabilities

	Status		
Graduate	Enlisted	Officer	Total
Selected	0.922807[a]	0.077193	1.000000
Non-selected	0.927273	0.072727	1.000000
Total	0.926000[a]	0.074000	1.000000

Note: These values represent the *likelihood* for Status given Graduate, for each event
[a] From Table 5.46, 0.922807 $=$ 263/285

Table 5.50 Likelihood probabilities

		PT			
Status	Graduate	Above	Extreme	Average	Total
Enlisted	Selected	0.349810[a]	0.311787	0.338403	1.000000
Enlisted	Non-selected	0.357466	0.352941	0.289593	1.000000
Officer	Selected	0.409091	0.272727	0.318182	1.000000
Officer	Non-selected	0.346154	0.307692	0.346154	1.000000

Note: These values represent the *likelihood* for PT given Status and Graduate for each event
[a] From Table 5.47, 0.349810 $=$ 92/263

Step 7a: Determine likelihood probabilities. We compute *likelihoods* conditioned, on the conditional counts, in Table 5.46. To determine these percentages, we calculate probabilities across the observable event by dividing them with the total count of each unobservable event. We report these results in Table 5.49.

Step 7b: Determine likelihood probabilities. We compute *likelihoods* conditioned, on the conditional counts, in Table 5.47. To determine these percentages, we calculate probabilities across the observable event by dividing them with the total count of each unobservable event. We report these results in Table 5.49.

Step 8: Determine joint and marginal probabilities. To compute the *joints*, we multiply the *priors* times the *likelihoods* in Tables 5.48, 5.49, and 5.50, across the events of Status, Graduate and PT. To compute the *marginals* of Status and Graduate events, we sum across the joint events of PT, and to compute those of PT, we sum down the events of Status and Graduate. The sum of the individual *joints* totals 100.0 %. We report the results in Table 5.51.

Table 5.51 Joint and marginal probabilities

Status	Graduate	PT Above	Extreme	Average	Marginal
Enlisted	Selected	0.092000[a]	0.082000	0.089000	0.263000
Enlisted	Non-selected	0.237000	0.234000	0.192000	0.663000
Officer	Selected	0.009000	0.006000	0.007000	0.022000
Officer	Non-selected	0.018000	0.016000	0.018000	0.052000
Marginal		0.356000	0.338000	0.306000	1.000000

Note: These values represent the *joints* for each PT, Status, and Graduate event, that we calculated using the *priors* and *likelihoods*
[a] From Tables 5.48, 5.49, and 5.50, 0.092000 = 0.285000×0.922807×0.349810

Table 5.52 Posterior probabilities

Status	PT	Graduate Selected	Non-selected	Total
Enlisted	Above	0.279635[a]	0.720365	1.000000
Enlisted	Extreme	0.259494	0.740506	1.000000
Enlisted	Average	0.316726	0.683274	1.000000
Officer	Above	0.333333	0.666667	1.000000
Officer	Extreme	0.272727	0.727273	1.000000
Officer	Average	0.280000	0.720000	1.000000

[a]From Table 5.51, 0.279635 = 0.092000/(0.092000 + 0.237000). This represents the probability of Graduate = Selected given Status = Enlisted and PT = Above

Table 5.53 Posterior probabilities

PT	Graduate	Status Enlisted	Officer	Total
Above	Selected	0.910891[a]	0.089109	1.000000
Extreme	Non-selected	0.936000	0.064000	1.000000
Average	Selected	0.927083	0.072917	1.000000
Above	Non-selected	0.929412	0.070588	1.000000
Extreme	Selected	0.931818	0.068182	1.000000
Average	Non-selected	0.914286	0.085714	1.000000

[a]From Table 5.51, 0.910891 = 0.092000/(0.092000 + 0.009000). This represents the probability of Status = Enlisted given PT = Above and Graduate = Selected

Step 9: Determine posterior probabilities. We compute *posteriors*, based on the *joints and marginals* in Table 5.51. To determine these probabilities, we divide the Status and PT joint event by the respective Graduate marginal event. We report the results in Table 5.52.

We compute the *posteriors* based on the *joints and marginals* in Table 5.51. To determine these probabilities, we divide the Status and PT joint event by the respective Graduate marginal event.

References

Grover, J. (2013). Strategic economic decision-making: Using Bayesian belief networks to solve complex problems. In: *Springer briefs in statistics* (Vol. 9). New York: Springer Science+Business Media. doi:10:1007/978-1-4614-6040-41.

Rickard, J. (2012). *Currency wars: The making of the next global crisis*. New York: Penguin.

Special Operations Recruiting. (n.d.). Retrieved November 06, 2016, from http://sorbrecruiting.com/

Weiers, R. M., Gray, B. J., & Peters, L. H. (2005). *Introduction to business statistics* (5th ed.). Mason, OH: South-Western Cengage Learning.

WELCOME TO OANDA. (n.d.). Retrieved November 06, 2016, from https://www.oanda.com/

Chapter 6
Base Matrices

6.1 Introduction

This chapter represents the statistical methodology we will follow in building and invoking a series of 2, 3, and 4 node Bayesian Belief Networks (BBN). We will demonstrate the steps using the concepts of statistics and probability theory developed in Chaps. 1 through 4.[1] We will first conduct model fitting to determine the number of joint and disjoint BBN nodal combinations and then create synthetic BBN to determine resultant *posterior* probabilities.

6.2 Model Fitting

To model these BBN, we will develop a series of two to four synthetics BBN with nodes containing one to six paths. We will create each node using frequency counts from a random experiment, and then convert them into *likelihoods*. The idea is that when we specify the BBN the starting point of reference is always the unobservable node *A*, which contains the *priors*, followed by appropriate *likelihood* and *conditional* probabilities. We combine them using the chain rule (Sect. 3.8) to create *joint* events, and then finally compute and report the resultant *posteriors*. These *joints* then become the building blocks for the *posteriors*.

To fit these BBN, we will first transition from variables representing columns of data into BBN nodes. Then we will specify the structure of these node pairs and combinations of Nodes *A*, *B*, *C*, & *D*, which are either joint or disjoint,[2] and string together their unique combinations based on their respective structures. The parent

---

[1]Chapter 5 was a rewrite of Grover, 2013.

[2]Joint nodes are those that have common elements, and disjoint nodes have no common elements.

© Springer International Publishing AG 2016
J. Grover, *The Manual of Strategic Economic Decision Making*,
DOI 10.1007/978-3-319-48414-3_6

node will always be A, followed by one or more children nodes, depending on the model fit. For each BBN, we can determine the number of nodal combinations using Eq. (3.4).[3]

6.2.1   2-Node BBN for Variables A & B

6.2.1.1   2-Node | 1-Path BBN

This BBN has one path as calculated in Eq. (6.1), which is a fully-specified BBN:

$$Path_n = \frac{1!}{(1-1)! * 1!} = 1, \qquad (6.1)$$

where n = 1, r = 1, and using node pairs: {AB}. The single path here is:

- Path 1: $[A \rightarrow B]$

6.2.2   3-Node BBN for Variables A, B, & C

6.2.2.1   3-Node | 2-Path BBN

This BBN has three path combinations, as calculated in Eq. (6.2):

$$Path_n = \frac{3!}{(3-2)! * 2!} = 3, \qquad (6.2)$$

where n = 3, r = 2, and using node: {AB, AC, BC}. The three paths here include:

- Path 1: $[A \rightarrow B | A \rightarrow C]$
- Path 2: $[A \rightarrow B | B \rightarrow C]$
- Path 3: $[A \rightarrow C | B \rightarrow C]$

6.2.3   3-Node | 3-Path BBN

This BBN has one path, as calculated in Eq. (6.3), which is a fully-specified BBN:

$$Path_n = \frac{3!}{(3-3)! * 3!} = 1, \qquad (6.3)$$

---

[3]We used the online combinations calculator @ http://www.mathsisfun.com/combinatorics/combinations-permutations-calculator.html. We calculated the number of paths for each BBN using Eq. (3.4) and referenced the values of n and r.

where n = 3, r = 3, and using node pairs: $\{AB, AC, BC\}$. This is the path:

- Path 1: $[A \rightarrow B | A \rightarrow C | B \rightarrow C]$

6.2.4   4-Node BBN for Variables A, B, C, & D

6.2.4.1   4-Node | 3-Path BBN

This BBN has 20 path combinations, as calculated in Eq. (6.4):

$$Path_n = \frac{6!}{(6-3)! * 3!} = 20, \tag{6.4}$$

where n = 6, r = 3 and using node pairs: $\{AB, AC, AD, BC, BD, CD\}$. The 20 paths[4] here include:

- Path 1: $[A \rightarrow B | A \rightarrow C | A \rightarrow D]$
- *Path 2: $[A \rightarrow B | A \rightarrow C | B \rightarrow C]$
- Path 3: $[A \rightarrow B | A \rightarrow C | B \rightarrow D]$
- Path 4: $[A \rightarrow B | A \rightarrow C | C \rightarrow D]$
- Path 5: $[A \rightarrow B | A \rightarrow D | B \rightarrow C]$
- *Path 6: $[A \rightarrow B | A \rightarrow D | B \rightarrow D]$
- Path 7: $[A \rightarrow B | A \rightarrow D | C \rightarrow D]$
- Path 8: $[A \rightarrow B | B \rightarrow C | B \rightarrow D]$
- Path 9: $[A \rightarrow B | B \rightarrow C | C \rightarrow D]$
- Path 10: $[A \rightarrow B | B \rightarrow D | C \rightarrow D]$
- Path 11: $[A \rightarrow C | A \rightarrow D | B \rightarrow C]$
- Path 12: $[A \rightarrow C | A \rightarrow D | B \rightarrow D]$
- *Path 13: $[A \rightarrow C | A \rightarrow D | C \rightarrow D]$
- Path 14: $[A \rightarrow C | B \rightarrow C | B \rightarrow D]$
- Path 15: $[A \rightarrow C | B \rightarrow C | C \rightarrow D]$
- Path 16: $[A \rightarrow C | B \rightarrow D | C \rightarrow D]$
- Path 17: $[A \rightarrow D | B \rightarrow C | B \rightarrow D]$
- Path 18: $[A \rightarrow D | B \rightarrow C | C \rightarrow D]$
- Path 19: $[A \rightarrow D | B \rightarrow D | C \rightarrow D]$
- *Path 20: $[B \rightarrow C | B \rightarrow D | C \rightarrow D]$

---

[4]Note: * Denotes randomly omitted paths.

6.2.5   4-Node | 4-Path BBN

This BBN has 15 path combinations, as calculated in Eq. (6.5):

$$Path_n = \frac{6!}{(6-4)! * 4!} = 15, \tag{6.5}$$

where n = 6, r = 4, and using node pairs: $\{AB, AC, AD, BC, BD, CD\}$. The 15 paths here include:

- Path 1: $[A \rightarrow B | A \rightarrow C | A \rightarrow D | B \rightarrow C]$
- Path 2: $[A \rightarrow B | A \rightarrow C | A \rightarrow D | B \rightarrow D]$
- Path 3: $[A \rightarrow B | A \rightarrow C | A \rightarrow D | C \rightarrow D]$
- Path 4: $[A \rightarrow B | A \rightarrow C | B \rightarrow C | B \rightarrow D]$
- Path 5: $[A \rightarrow B | A \rightarrow C | B \rightarrow C | C \rightarrow D]$
- Path 6: $[A \rightarrow B | A \rightarrow C | B \rightarrow D | C \rightarrow D]$
- Path 7: $[A \rightarrow B | A \rightarrow D | B \rightarrow C | B \rightarrow D]$
- Path 8: $[A \rightarrow B | A \rightarrow D | B \rightarrow C | C \rightarrow D]$
- Path 9: $[A \rightarrow B | A \rightarrow D | B \rightarrow D | C \rightarrow D]$
- Path 10: $[A \rightarrow B | B \rightarrow C | B \rightarrow D | C \rightarrow D]$
- Path 11: $[A \rightarrow C | A \rightarrow D | B \rightarrow C | B \rightarrow D]$
- Path 12: $[A \rightarrow C | A \rightarrow D | B \rightarrow C | C \rightarrow D]$
- Path 13: $[A \rightarrow C | A \rightarrow D | B \rightarrow D | C \rightarrow D]$
- Path 14: $[A \rightarrow C | B \rightarrow C | B \rightarrow D | C \rightarrow D]$
- Path 15: $[A \rightarrow D | B \rightarrow C | B \rightarrow D | C \rightarrow D]$

6.2.6   4-Node | 5-Path BBN

This BBN has six path combinations, as calculated in Eq. (6.6):

$$Path_n = \frac{6!}{(6-5)! * 5!} = 6, \tag{6.6}$$

where n = 6, r = 5, and using node pairs: $\{AB, AC, AD, BC, BD, CD\}$. The six paths here include:

- Path 1: $[A \rightarrow B | A \rightarrow C | A \rightarrow D | B \rightarrow C | B \rightarrow D]$
- Path 2: $[A \rightarrow B | A \rightarrow C | A \rightarrow D | B \rightarrow C | C \rightarrow D]$
- Path 3: $[A \rightarrow B | A \rightarrow C | A \rightarrow D | B \rightarrow D | C \rightarrow D]$
- Path 4: $[A \rightarrow B | A \rightarrow C | B \rightarrow C | B \rightarrow D | C \rightarrow D]$
- Path 5: $[A \rightarrow B | A \rightarrow D | B \rightarrow C | B \rightarrow D | C \rightarrow D]$
- Path 6: $[A \rightarrow C | A \rightarrow D | B \rightarrow C | B \rightarrow D | C \rightarrow D]$

6.2.7   4-Node | 6-Path BBN

This BBN has one path as calculated in Eq. (6.7), which is a fully-specified BBN:

$$Path_n = \frac{6!}{(6-6)! * 6!} = 1, \qquad (6.7)$$

where n = 6, r = 6, and using node pairs: $\{AB, AC, AD, BC, BD, CD\}$. The single path here is:

- Path 1: $[A \to B|A \to C|A \to D|B \to C|B \to D|C \to D]$

6.3   Synthetic Node Creation

In Sect. 6.2, we established the number of unique paths for BBN with 2 to 4 nodes from the following subspace of nodes: $\{A, B, C, D\}$. We will create synthetics nodes using Node A as our base to create our framework for BBN creation in Chaps. 7 through 9. Each synthetic will have one joint and a series of disjoint structure/s based on these combinations in Sect. 6.2. Once we determine these paths, we will create our series of synthetics from each nodal combination. We will separate each Node, A, B, C, and D into joint and disjoint nodes and represent the disjoints using strikethroughs in the likelihood probabilities.

Table 6.1 computes the joint counts and Table 6.2 calculates the joint event sample space of conditional counts, which is the base for calculating the *likelihood* probabilities of each synthetic node.[5]

6.3.1   Random Experimental Results

Using the statistical definitions defined in Chap. 3, we will create a dataset using Monte Carlo simulations as our sample space containing the four random variables, A, B, C, and D, each with two events, True and False. We will transition these variables into nodes as we create the BBN. We will create each node with different proportions and based on our random experiment of 10,000 trials. This becomes our subset (Sect. 3.32.2) of data for the BBN. Table 6.1 reports the total (deflated) frequency counts for each node-event combinations.

Table 6.2 reports the total marginal probabilities of each node.

---

[5]We counted and populated these data, from the random experiment for this manual.

Table 6.1 Total variable frequency counts

Variable (Node)				
Event	A	B	C	D
True	1023[a]	3029	5005	7134
False	8977	6971	4995	2866
Total	10,000[b]	10,000	10,000	10,000

Note: These values represent the total variable frequency counts obtained from the sampling process
[a] 1023 represents the count of all True events in variable *A*
[b] Represents the total number of trials

Table 6.2 Total variable marginal probabilities

Variable (Node)				
Event	A	B	C	D
True	0.102300[a]	0.302900	0.500500	0.713400
False	0.897700	0.697100	0.499500	0.286600
Total	1.000000	1.000000	1.000000	1.000000

Note: These values represent the total variable frequency counts obtained from the sampling process
[a] From Table 6.1, 0.102300 = 1023/10,000 and represents the *marginal* probability of all True events in variable *A*

6.3.1.1   Random Experimental Results-Expanded

Table 6.3 reports the joint space counts for the event combinations (True & False) for these nodes. These become the base for calculating the *likelihoods* of each synthetic node.[6]

Table 6.4 reports the joint space counts for the event combinations (True & False) of these respective nodes.

6.4   Node *A*

6.4.1   Node Determination

Node *A* is the only one used for the prior probabilities.

---

[6] These data are counted from the random experiment conducted to populate the data for this manual.

Table 6.3 Expanded variable joint event frequency counts

$P(D1_j, C1_i, B1_i, A_i)$					
$D1_j$					
$C1_i$	$B1_i$	A_i	True	False	Marginals
True	True	True	88[a]	50	138
True	True	False	923	398	1321
True	False	True	260	88	348
True	False	False	2285	913	3198
False	True	True	115	43	158
False	True	False	1009	403	1412
False	False	True	284	95	379
False	False	False	2170	876	3046
Marginals			7134	2866	10,000

[a]$88 = C(D1_{True}, C1_{True}, B1_{True}, A_{True})$ as counted from the random experiment conducted to populate the data for this experiment. There are 88 True counts in this joint event. This is a precursor to computing the *likelihoods*

Table 6.4 Joint and marginal probabilities

$P(D1_j, C1_i, B1_i, A_i)$					
$D1_j$					
$C1_i$	$B1_i$	A_i	True	False	Marginals
True	True	True	0.008800[a]	0.005000	0.013800
True	True	False	0.092300	0.039800	0.132100
True	False	True	0.026000	0.008800	0.034800
True	False	False	0.228500	0.091300	0.319800
False	True	True	0.011500	0.004300	0.015800
False	True	False	0.100900	0.040300	0.141200
False	False	True	0.028400	0.009500	0.037900
False	False	False	0.217000	0.087600	0.304600
Marginals			0.713400	0.286600	1.000000

[a]From Table 6.3, $0.008800 = 88/10,000$

6.4.2   Frequency Counts and Likelihood Probabilities

6.4.2.1   Node A: Frequency Counts

Table 6.5 reports the total (deflated) frequency counts for Node A, which is the only synthetic node in this series, and it will always represent the prior probabilities.

6.4.2.2   Node A: Prior Probabilities

Table 6.6 reports the *priors* for Node A based on the frequency counts from Table 6.5. We will use this node as our starting reference point for each BBN.

Table 6.5 Node *A*:
frequency counts

C(A)		
True	False	Total
1023[a]	8977	10,000

[a]From the marginal counts of Table 6.3, $1023 = 138 + 348 + 158 + 379$

Table 6.6 Node *A*: prior
probabilities

$P(A_j)$		
True	False	Total
0.102300[a]	0.897700	1.000000

[a]From Table 6.5, $0.102300 = 1023/10{,}000$

Table 6.7 Node B synthetics

B	A
B1	A
B2	~~A~~

6.5 Node B Synthetics

In this section and henceforth, we will see strikethroughs of disjoint nodes. We have marginalized them out because they have no effect on the conditional probability results.

6.5.1 Node Determination

To determine the number of synthetic nodes, Table 6.7 reports the combinations of events, from each node we identified in Sect. 6.2.2.1 for Node(s) *A*, and *B1 & B2*.[7]

6.5.2 Frequency Counts and Likelihood Probabilities

6.5.2.1 Node *B*1: Joint Space Frequency Counts

Table 6.8 reports the total (deflated) frequency counts for Node *B1*.

---

[7]The strikethrough represent a disjoint node.

6.5.2.2 Node *B1*: Likelihood Probabilities

Table 6.9 reports the *likelihoods* for Node *B1*, based on the frequency counts from Table 6.8.

6.5.2.3 Node *B2*: Joint Space Frequency Counts

Table 6.10 reports the total (deflated) frequency counts for Node *B2*.

6.5.2.4 Node *B2*: Likelihood Probabilities

Table 6.11 reports the *likelihoods* for Node *B2*, based on the frequency counts from Table 6.10, where Node *A* is disjoint.

6.6 Node C Synthetics

6.6.1 Node Determination

To determine the number of synthetic nodes, Table 6.12 reports the combinations of events from each node we identified in Sect. 6.2.3, for node(s) *A*, and *B1 & B2*, and *C1, C2, C3, & C4*.[8]

Table 6.8 Node *B1*: frequency counts

$C(B1_j, A_i)$

$B1_j$			
A_i	True	False	Total
True	296[a]	727	1023
False	2733	6244	8977
Total	3029	6971	10,000

[a]From the marginal counts of Table 6.3, $296 = 138 + 158$

Table 6.9 Node *B1*: likelihood probabilities

$P(B1_j|A_i)$

$B1_j$			
A_i	True	False	Total
True	0.289345[a]	0.710655	1.000000
False	0.304445	0.695555	1.000000

[a]From Table 6.8, $C(B1_{True}|A_{True})$, $0.289345 = 296/1023$

---

[8]Strikethroughs represent disjoint nodes.

Table 6.10 Node $B2$: frequency counts

$P(B2_j, A_i)$			
$B2_j$			
A_i	True	False	Total
True	3029[a]	6971	10,000
False	3029	6971	10,000
Total	3029	6971	10,000

[a]From the marginal counts of Table 6.3, $3029 = 138 + 1321 + 158 + 1412$

Table 6.11 Node $B2$: likelihood probabilities

$P(B2_j|A_i)$			
$B2_j$			
A_i	True	False	Total
~~True~~	0.302900[a]	0.697100	1.000000
~~False~~	~~0.302900~~	~~0.697100~~	~~1.000000~~

[a]From Table 6.10, $C(B2_{True}|A_{True}) = C(B2_{True})$, $0.302900 = 3029/10{,}000$

Table 6.12 Node C synthetics

C	B	A
C1	B	A
C2	B	~~A~~
C3	~~B~~	A
C4	~~B~~	~~A~~

Table 6.13 Node $C1$: frequency counts

$C(C1_j, B_i, A_i)$				
$C1_j$				
B_i	A_i	True	False	Total
True	True	138[a]	158	296
True	False	1321	1412	2733
False	True	348	379	727
False	False	3198	3046	6244
Total		5005	4995	10,000

[a]From the marginal counts of Table 6.3, $138 = 88 + 50$

6.6.2 Frequency Counts and Likelihood Probabilities

6.6.2.1 Node C1: Joint Space Frequency Counts

Table 6.13 reports the total (deflated) frequency counts for Node $C1$.

Table 6.14 Node $C1$: likelihood probabilities

$P(C1_j|B_i, A_i)$				
$C1_j$				
B_i	A_i	True	False	Total
True	True	0.466216[a]	0.533784	1.000000
True	False	0.483352	0.516648	1.000000
False	True	0.478680	0.521320	1.000000
False	False	0.512172	0.487828	1.000000

[a]From Table 6.13, $P(C1_{True}|B_{True}, A_{True})$, $0.466216 = 138/296$

Table 6.15 Node $C2$: frequency counts

$C(C2_j, B_i, A_i)$				
$C2_j$				
B_i	A_i	True	False	Total
True	True	1459[a]	1570	3029
True	False	1459	1570	3029
False	True	3546	3425	6971
True	False	3546	3425	6971
Total		5005	4995	10,000

[a]From the marginal counts of Table 6.3, $1459 = 138 + 1321$

6.6.2.2 Node $C1$: Likelihood Probabilities

Table 6.14 reports the *likelihoods* for Node $C1$, based on the frequency counts from Table 6.13.

6.6.2.3 Node $C2$: Joint Space Frequency Counts

Table 6.15 reports the total (deflated) frequency counts for Node $C2$.

6.6.2.4 Node $C2$: Likelihood Probabilities

Table 6.16 reports the *likelihoods* for Node $C2$, based on the frequency counts from Table 6.15, where Node A is disjoint.

6.6.2.5 Node $C3$: Joint Space Frequency Counts

Table 6.17 reports the total (deflated) frequency counts for Node $C3$, where Nodes B is independent.

Table 6.16 Node $C2$: likelihood probabilities

$P(C2_j\|B_i, A_i)$				
$C2_j$				
B_i	A_i	True	False	Total
True	True	0.481677[a]	0.518323	1.000000
True	False	0.481677	0.518323	1.000000
False	True	0.508679	0.491321	1.000000
False	False	0.508679	0.491321	1.000000

[a]From Table 6.15, $P(C2_{True}|B_{True}, A_{True})$ $P(C2_{True}|B_{True}), 0.481677 = 1459/3029$

Table 6.17 Node $C3$: frequency counts

$C(C3_j, B_i, A_i)$				
$C3_j$				
B_i	A_i	True	False	Total
True	True	486[a]	537	1023
True	False	4519	4458	8977
False	True	486	537	1023
False	False	4519	4458	8977
Total		5005	4995	10,000

[a]From the marginal counts of Table 6.3, $486 = 138 + 348$

Table 6.18 Node $C3$: likelihood probabilities

$P(C3_j\|B_i, A_i)$				
$C3_j$				
B_i	A_i	True	False	Total
True	True	0.475073[a]	0.524927	1.000000
True	False	0.503398	0.496602	1.000000
False	True	0.475073	0.524927	1.000000
False	False	0.503398	0.496602	1.000000

[a]From Table 6.17, $C(C3_{True}|B_{True}, A_{True})$ = $C(C3_{True}|A_{True}) = 0.475073 = 486/1023$

6.6.2.6 Node $C3$: Likelihood Probabilities

Table 6.18 reports the *likelihoods* for Node $C3$, based on the frequency counts from Table 6.17, where Node $B1$ is disjoint.

6.6.2.7 Node $C4$: Joint Space Frequency Counts

Table 6.19 reports the total (deflated) frequency counts for Node $C4$.

Table 6.19 Node $C4$: frequency counts

$C(C4_j, B_i, A_i)$				
$C4_j$				
B_i	A_i	True	False	Total
True	True	5005[a]	4995	10,000
True	False	5005	4995	10,000
False	True	5005	4995	10,000
False	False	5005	4995	10,000
Total		5005	4995	10,000

[a]From the marginal counts of Table 6.3, $5005 = 138 + 1321 + 348 + 3198$

Table 6.20 Node $C4$: likelihood probabilities

$P(C4_j \mid B_i, A_i)$				
$C4_j$				
B_i	A_i	True	False	Total
~~True~~	~~True~~	0.500500[a]	0.499500	1.000000
~~True~~	~~False~~	0.500500	0.499500	1.000000
~~False~~	~~True~~	0.500500	0.499500	1.000000
~~False~~	~~False~~	0.500500	0.499500	1.000000

[a]From Table 6.19, $P(C4_{True} \mid \cancel{B_{True}}, \cancel{A_{True}}) = P(C4_{True}) = 0.500500 = 5005/10,000$

6.6.2.8 Node $C4$: Likelihood Probabilities

Table 6.20 reports the *likelihoods* for Node $C4$, based on the frequency counts from Table 6.19, where Nodes B2 and A are disjoint.

6.7 Node D Synthetics

6.7.1 Node Determination

To determine the number of synthetic nodes, Table 6.21 reports the combinations of events from each node, we identified in Chap. 6.2.4, for node(s) A, and $B1$ & $B2$, and $C1$, $C2$, $C3$, & $C4$, and $D1$, $D2$, $D3$, $D4$, $D5$, $D6$, & $D7$,[9]

---

[9]Strikethroughs represent disjoint nodes.

Table 6.21 Node D
synthetics

D_j	C_j	B_j	A_j
D1	C	B	A
D2	C	B_i	~~A~~
D3	C	~~B~~	A
D4	~~C~~	B	A
D5	C	~~B~~	~~A~~
D6	~~C~~	B	~~A~~
D7	~~C~~	~~B~~	A
D8[a]	~~C~~	~~B~~	~~A~~

[a]This node is not required and was not included

Table 6.22 Node $D1$:
frequency counts

$C(D1_j, C_i, B_i, A_i)$						
$D1_j$						
C_i	B_i	A_i	True	False	Marginals	
True	True	True	88[a]	50	138	
True	True	False	923	398	1321	
True	False	True	260	88	348	
True	False	False	2285	913	3198	
False	True	True	115	43	158	
False	True	False	1009	403	1412	
False	False	True	284	95	379	
False	False	False	2170	876	3046	
Marginals			7134	2866	10,000	

[a]From the *joint* counts of Table 6.3

6.7.2 Frequency Counts and Likelihood Probabilities

6.7.2.1 Node $D1$: Joint Space Frequency Counts

Table 6.22 reports the total (deflated) frequency counts for Node $D1$.

6.7.2.2 Node $D1$: Likelihood Probabilities

Table 6.23 reports the *likelihoods* for Node $D1$, based on the frequency counts from Table 6.22.

6.7.2.3 Node $D2$: Joint Space Frequency Counts

Table 6.24 reports the total (deflated) frequency counts for Node $D2$.

Table 6.23 Node $D1$:
likelihood probabilities

$P(D1_j|C_i, B_i, A_i)$

$D1_j$

C_i	B_i	A_i	True	False	Total
True	True	True	0.637681[a]	0.362319	1.000000
True	True	False	0.698713	0.301287	1.000000
True	False	True	0.747126	0.252874	1.000000
True	False	False	0.714509	0.285491	1.000000
False	True	True	0.727848	0.272152	1.000000
False	True	False	0.714589	0.285411	1.000000
False	False	True	0.749340	0.250660	1.000000
False	False	False	0.712410	0.287590	1.000000

[a]From Table 6.22, $P(D1_{True}|C_{True}, B_{True}, A_{True})$, $0.637681 = 88/138$

Table 6.24 Node $D2$:
frequency counts

$C(D1_j, C_i, B_i, A_i)$

$D1_j$

C_i	B_i	A_i	True	False	Total
True	True	True	1011[a]	448	1459
True	True	False	1011	448	1459
True	False	True	2545	1001	3546
True	False	False	2545	1001	3546
False	True	True	1124	446	1570
False	True	False	1124	446	1570
False	False	True	2454	971	3425
False	False	False	2454	971	3425
Total			7134	2866	10,000

[a]From the marginal counts of Table 6.3, $1011 = 88 + 923$

6.7.2.4   Node $D2$: Likelihood Probabilities

Table 6.25 reports the *likelihoods* for Node $D2$, based on the frequency counts from Table 6.24.

6.7.2.5   Node $D3$: Joint Space Frequency Counts

Table 6.26 reports the total (deflated) frequency counts for Node $D3$.

6.7.2.6   Node $D3$: Likelihood Probabilities

Table 6.27 reports the *likelihoods* for Node $D3$, based on the frequency counts from Table 6.26.

Table 6.25 Node $D2$: likelihood probabilities

$P(D2_j\|C_i, B_i, A_i)$						
$D2_j$						
C_i	B_i	A_i	True	False	Total	
True	True	True	0.692940[a]	0.307060	1.000000	
~~True~~	~~True~~	~~False~~	~~0.692940~~	~~0.307060~~	~~1.000000~~	
True	False	True	0.717710	0.282290	1.000000	
~~True~~	~~False~~	~~False~~	~~0.717710~~	~~0.282290~~	~~1.000000~~	
False	True	True	0.715924	0.284076	1.000000	
~~False~~	~~True~~	~~False~~	~~0.715924~~	~~0.284076~~	~~1.000000~~	
False	False	True	0.716496	0.283504	1.000000	
~~False~~	~~False~~	~~False~~	~~0.716496~~	~~0.283504~~	~~1.000000~~	

[a]From Table 6.24, $P(D2_{True}|C_{True}, B_{True}, A_{True}) = P(D2_{True}|C_{True}, B_{True}) = 0.692940 = 1011/1459$

Table 6.26 Node $D3$: frequency counts

$C(D3_j, C_i, B_i, A_i)$					
$D3_j$					
C_i	B_i	A_i	True	False	Total
True	True	True	348[a]	138	486
True	True	False	3208	1311	4519
True	False	True	348	138	486
True	False	False	3208	1311	4519
False	True	True	399	138	537
False	True	False	3179	1279	4458
False	False	True	399	138	537
False	False	False	3179	1279	4458
Totals			7134	2866	10,000

[a]From the *joint* counts of Table 6.3, $348 = 88 + 260$

Table 6.27 Node $D3$: likelihood probabilities

$P(D3_j\|C_i, B_i, A_i)$						
$D3_j$						
C_i	B_i	A_i	True	False	Total	
True	True	True	0.716049[a]	0.283951	1.000000	
True	True	False	0.709892	0.290108	1.000000	
~~True~~	~~False~~	~~True~~	~~0.716049~~	~~0.283951~~	~~1.000000~~	
~~True~~	~~False~~	~~False~~	~~0.709892~~	~~0.290108~~	~~1.000000~~	
False	True	True	0.743017	0.256983	1.000000	
False	True	False	0.713100	0.286900	1.000000	
~~False~~	~~False~~	~~True~~	~~0.743017~~	~~0.256983~~	~~1.000000~~	
~~False~~	~~False~~	~~False~~	~~0.713100~~	~~0.286900~~	~~1.000000~~	

[a]From Table 6.26, $P(D3_{True}|C_{True}, B_{True}, A_{True}) = P(D3_{True}|C_{True}, A_{True}) = 0.716049 = 348/486$

Table 6.28 Node $D4$: frequency counts

$P(D4_j, C_i, B_i, A_i)$					
$D4_j$					
C_i	B_i	A_i	True	False	Total
True	True	True	203[a]	93	296
True	True	False	1932	801	2733
True	False	True	544	183	727
True	False	False	4455	1789	6244
False	True	True	203	93	296
False	True	False	1932	801	2733
False	False	True	544	183	727
False	False	False	4455	1789	6244
Totals			7134	2866	10,000

[a]From the *joint* counts of Table 6.3, $203 = 88 + 115$

Table 6.29 Node $D4$: likelihood probabilities

$P(D4_j \mid \cancel{C_i}, B_i, A_i)$					
$D4_j$					
$\cancel{C_i}$	B_i	A_i	True	False	Total
~~True~~	True	True	0.685811[a]	0.314189	1.000000
~~True~~	True	False	0.706915	0.293085	1.000000
~~True~~	False	True	0.748281	0.251719	1.000000
~~True~~	False	False	0.713485	0.286515	1.000000
~~False~~	~~True~~	~~True~~	~~0.685811~~	~~0.314189~~	~~1.000000~~
~~False~~	~~True~~	~~False~~	~~0.706915~~	~~0.293085~~	~~1.000000~~
~~False~~	~~False~~	~~True~~	~~0.748281~~	~~0.251719~~	~~1.000000~~
~~False~~	~~False~~	~~False~~	~~0.713485~~	~~0.286515~~	~~1.000000~~

[a]From Table 6.28, $P(D4_{True} \mid \cancel{C_{True}}, B_{True}, A_{True}) = P(D4_{True} \mid B_{True}, A_{True}) = 0.685811 = 203/296$

6.7.2.7 Node $D4$: Joint Space Frequency Counts

Table 6.28 reports the total (deflated) frequency counts for Node $D4$.

6.7.2.8 Node $D4$: Likelihood Probabilities

Table 6.29 reports the *likelihoods* for Node $D4$, based on the frequency counts from Table 6.28.

6.7.2.9 Node $D5$: Joint Space Frequency Counts

Table 6.30 reports the total (deflated) frequency counts for Node $D5$.

Table 6.30 Node $D5$:
frequency counts

$C(D5_j, C_i, B_i, A_i)$					
$D5_j$					
C_i	B_i	A_i	True	False	Total
True	True	True	3556[a]	1449	5005
True	True	False	3556	1449	5005
True	False	True	3556	1449	5005
True	False	False	3556	1449	5005
False	True	True	3578	1417	4995
False	True	False	3578	1417	4995
False	False	True	3578	1417	4995
False	False	False	3578	1417	4995
Totals			7134	2866	10,000

[a]From the *joint* counts of Table 6.3, $3556 = 88 + 923 + 260 + 2285$

Table 6.31 Node $D5$:
likelihood probabilities

$P(D5_j | C_i, B_i, A_i)$					
$D5_j$					
C_i	B_i	A_i	True	False	Total
True	True	True	0.710490[a]	0.289510	1.000000
True	True	False	0.710490	0.289510	1.000000
True	False	True	0.710490	0.289510	1.000000
True	False	False	0.710490	0.289510	1.000000
False	True	True	0.716316	0.283684	1.000000
False	True	False	0.716316	0.283684	1.000000
False	False	True	0.716316	0.283684	1.000000
False	False	False	0.716316	0.283684	1.000000

[a]From Table 6.30, $P(D5_{True}|C_{True}, B_{True}, A_{True}) = P(D5_{True}|C_{True}) = 0.710490 = 3556/5005$

6.7.2.10   Node $D5$: Likelihood Probabilities

Table 6.31 reports the *likelihoods* for Node $D5$, based on the frequency counts from Table 6.30.

6.7.2.11   Node $D6$: Joint Space Frequency Counts

Table 6.32 reports the total (deflated) frequency counts for Node $D6$.

6.7.2.12   Node $D6$: Likelihood Probabilities

Table 6.33 reports the *likelihoods* for Node $D6$, based on the frequency counts from Table 6.32.

Table 6.32 Node $D6$: frequency counts

$C(D6_j, C_i, B_i, A_i)$					
$D6_j$					
C_i	B_i	A_i	True	False	Total
True	True	True	2135[a]	894	3029
True	True	False	2135	894	3029
True	False	True	4999	1972	6971
True	False	False	4999	1972	6971
False	True	True	2135	894	3029
False	True	False	2135	894	3029
False	False	True	4999	1972	6971
False	False	False	4999	1972	6971
Totals			7134	2866	10,000

[a] From the *joint* counts of Table 6.3, $2135 = 88 + 923 + 115 + 1{,}009$

Table 6.33 Node $D6$: likelihood probabilities

$P(D6_j | C_i, B_i, A_i)$					
$D6_j$					
C_i	B_i	A_i	True	False	Total
True	True	True	0.704853[a]	0.295147	1.000000
True	True	False	0.704853	0.295147	1.000000
True	False	True	0.717114	0.282886	1.000000
True	False	False	0.717114	0.282886	1.000000
False	True	True	0.704853	0.295147	1.000000
False	True	False	0.704853	0.295147	1.000000
False	False	True	0.717114	0.282886	1.000000
False	False	False	0.717114	0.282886	1.000000

[a] From Table 6.32, $P(D6_{True} | C_{True}, B_{True}, A_{True}) = P(D6_{True} | B_{True}) = 0.704853 = 2135/3029$

6.7.2.13   Node $D7$: Joint Space Frequency Counts

Table 6.34 reports the total (deflated) frequency counts for Node $D7$.

6.7.2.14   Node $D7$: Likelihood Probabilities

Table 6.35 reports the *likelihoods* for Node $D7$, based on the frequency counts from Table 6.34.

Table 6.34 Node $D7$: frequency counts

$C(D7_j, C_i, B_i, A_i)$					
$D7_j$					
C_i	B_i	A_i	True	False	Total
True	True	True	747[a]	276	1023
True	True	False	6387	2590	8977
True	False	True	747	276	1023
True	False	False	6387	2590	8977
False	True	True	747	276	1023
False	True	False	6387	2590	8977
False	False	True	747	276	1023
False	False	False	6387	2590	8977
Totals			7134	2866	10000

[a]From the marginal counts of Table 6.3, $747 = 88 + 260 + 115 + 284$

Table 6.35 Node $D7$: likelihood probabilities

$P(D7_j | C_i, B_i, A_i)$					
$D7_j$					
C_i	B_i	A_i	True	False	Total
~~True~~	~~True~~	True	0.730205[a]	0.269795	1.000000
~~True~~	~~True~~	False	0.711485	0.288515	1.000000
~~True~~	~~False~~	~~True~~	~~0.730205~~	~~0.269795~~	~~1.000000~~
~~True~~	~~False~~	~~False~~	~~0.711485~~	~~0.288515~~	~~1.000000~~
~~False~~	~~True~~	~~True~~	~~0.730205~~	~~0.269795~~	~~1.000000~~
~~False~~	~~True~~	~~False~~	~~0.711485~~	~~0.288515~~	~~1.000000~~
~~False~~	~~False~~	~~True~~	~~0.730205~~	~~0.269795~~	~~1.000000~~
~~False~~	~~False~~	~~False~~	~~0.711485~~	~~0.288515~~	~~1.000000~~

[a]From Table 6.34, $P(D7_{True} | C_{True}, B_{True}, A_{True}) = C(D7_{True} | A_{True}) = 0.730205 = 747/1023$

Reference

Combinations and Permutations Calculator. (n.d.). Retrieved November 06, 2016, from http://www.mathsisfun.com/combinatorics/combinations-permutations-calculator.html.

Chapter 7
2-Node BBN

7.1 Introduction

This chapter represents specification and presentation on a 2-Node Bayesian Belief Network using the BBN solution protocol with counts and likelihood probabilities we created from synthetic node process developed in Chap. 6.

7.2 Proof

$$P(A|B) = \frac{P(B,A)}{P(B)} \tag{7.1}$$

$$P(A,B) = P(B,A) \tag{7.2}$$

$$P(A,B) = P(A|B)P(B) \tag{7.3}$$

$$P(B,A) = P(B|A)P(A) \tag{7.4}$$

$$P(A|B)P(B) = P(B|A)P(A) \tag{7.5}$$

$$P(A|B) = \frac{P(B|A)P(A)}{P(B)} \tag{7.6}$$

$$P(A|B) = \frac{P(B,A)}{P(B)}\,\square \tag{7.7}$$

7.3 BBN Solution Protocol

Using the established research protocol in Sect. 4.2, we will use the following steps to specify and solve each BBN.

© Springer International Publishing AG 2016
J. Grover, *The Manual of Strategic Economic Decision Making*,
DOI 10.1007/978-3-319-48414-3_7

- Step 2: Specify the BBN
- Step 6: Determine prior or unconditional probabilities
- Step 7: Determine likelihood probabilities
- Step 8: Determine joint and marginal probabilities
- Step 9: Compute posterior probabilities

7.4   1-Path BBN

7.4.1   Path 1: BBN Solution Protocol

Step 2: Specify the BBN–[$A \rightarrow B1$]

See Fig. 7.1.

Fig. 7.1 Path [$A \rightarrow B1$]

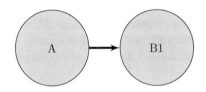

Step 6: Determine Prior or Unconditional Probabilities

We select Node A for this BBN structure as reported in Table 7.1.

Step 7: Determine Likelihood Probabilities

We select Node $B1$ for this BBN structure as reported in Table 7.2.

Step 8: Determine Joint and Marginal Probabilities

We compute the *joints* and *marginals* for $P(B1_j, A_i)$ and report these in Table 7.3.

Table 7.1 Node A: prior probabilities

$P(A_j)$		
True	False	Total
0.102300[a]	0.897700	1.000000

[a]From Table 6.5, $0.102300 = 1023/10{,}000$

Table 7.2 Node $B1$: likelihood probabilities

$P(B1_j\mid A_i)$			
$B1_j$			
A_i	True	False	Total
True	0.289345[a]	0.710655	1.000000
False	0.304445	0.695555	1.000000

[a]From Table 6.8, $0.289345 = 296/1023$

Table 7.3 Joint and marginal probabilities

$P(B1_j, A_i)$			
$B1_j$			
A_i	True	False	Marginals
True	0.029600[a]	0.072700	0.102300
False	0.273300	0.624400	0.897700
Marginals	0.302900	0.697100	1.000000

Note: ZSMS $= 0.000000 + 0.000000 = 0.000000$
[a]From Tables 7.1 and 7.2, $0.029600 = 0.289345 * 0.102300$

Table 7.4 Posterior probabilities

$P(A_j\mid B1_i)$			
A_j			
$B1_i$	True	False	Total
True	0.097722[a]	0.902278	1.000000
False	0.104289	0.895711	1.000000

[a]From Table 7.3, $0.097722 = 0.029600/0.302900$

Step 9: Determine Posterior Probabilities

We compute the *posteriors* for $P(A_j\mid B1_i)$ and report this in Table 7.4.

Chapter 8
3-Node BBN

8.1 Introduction

This chapter represents specification and presentation on a 3-Node Bayesian Belief Networks using the BBN solution protocol with counts and likelihood probabilities we created from the synthetic node process developed in Chap. 6.

8.2 Proof

$$P(A|B,C) = \frac{P(C,B,A)}{P(B,C)} \tag{8.1}$$

$$P(A,B,C) = P(C,B,A) \tag{8.2}$$

$$P(A,B,C) = P(A|B,C)P(B|C)P(C) \tag{8.3}$$

$$P(C,B,A) = P(C|B,A)P(B|A)P(A) \tag{8.4}$$

$$P(A|B,C)P(B|C)P(C) = P(C|B,A)P(B|A)P(A) \tag{8.5}$$

$$P(A|B,C) = \frac{P(C|B,A)P(B|A)P(A)}{P(B|C)P(C)} \tag{8.6}$$

$$P(A|B,C) = \frac{P(C|B,A)P(B,A)}{P(B|C)P(C)} \tag{8.7}$$

$$P(A|B,C) = \frac{P(C,B,A)}{P(B|C)P(C)} \tag{8.8}$$

$$P(A|B,C) = \frac{P(C,B,A)}{P(B,C)} \square \tag{8.9}$$

© Springer International Publishing AG 2016
J. Grover, *The Manual of Strategic Economic Decision Making*,
DOI 10.1007/978-3-319-48414-3_8

8.3   BBN Solution Protocol

- Step 2: Specify the BBN
- Step 6: Determine prior or *unconditional* probabilities
- Step 7: Determine likelihood probabilities
- Step 8: Determine joint and marginal probabilities
- Step 9: Compute posterior probabilities

8.4   2-Path BBN

8.4.1   Path 1: BBN Solution Protocol

Step 2: Specify the BBN–[$A \rightarrow C3|A \rightarrow B1$]

See Fig. 8.1.

Step 6: Determine Prior or Unconditional Probabilities

We select Node *A* for this BBN structure as reported in Table 8.1.

Fig. 8.1 Paths
[$A \rightarrow C3|A \rightarrow B1$]

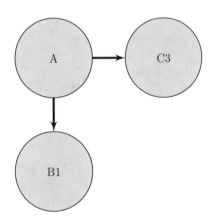

Table 8.1 Node *A*: prior
probabilities

$P(A_j)$		
True	False	Total
0.102300[a]	0.897700	1.000000

[a]From   Table   6.5,   0.102300   =
1023/10,000

Table 8.2 Node $B1$:
likelihood probabilities

$P(B1_j|A_i)$			
$B1_j$			
A_i	True	False	Total
True	0.289345[a]	0.710655	1.000000
False	0.304445	0.695555	1.000000

[a]From Table 6.8, 0.289345 = 296/1023

Table 8.3 Node $C3$:
likelihood probabilities

$P(C3_j|B1_i, A_i)$				
$C3_j$				
~~$B1_i$~~	A_i	True	False	Total
~~True~~	True	0.475073[a]	0.524927	1.000000
~~True~~	False	0.503398	0.496602	1.000000
~~False~~	~~True~~	~~0.475073~~	~~0.524927~~	~~1.000000~~
~~False~~	~~False~~	~~0.503398~~	~~0.496602~~	~~1.000000~~

[a]From Table 6.17, 0.475073 = 486/1023

Table 8.4 Joint and marginal
probabilities

$P(C3_j, B1_i, A_i)$				
$C3_j$				
$B1_i$	A_i	True	False	Marginals
True	True	0.014062[a]	0.015538	0.029600
True	False	0.137579	0.135721	0.273300
False	True	0.034538	0.038162	0.072700
False	False	0.314321	0.310079	0.624400
Marginals		0.500500	0.499500	1.000000

Note: ZSMS = 0.000000 + 0.000000 = 0.000000
[a]From Tables 8.1, 8.2, and 8.3, 0.014062 = 0.475073 * 0.289345 * 0.102300

Step 7: Determine Likelihood Probabilities

We select Node $B1$ for this BBN structure as reported in Table 8.2.
We select Node $C3$ for this BBN structure as reported in Table 8.3.

Step 8: Determine Joint and Marginal Probabilities

We compute the *joints& marginals* for $P(C3_j, B1_i, A_i)$ and report these in Table 8.4.

Step 9: Compute Posterior Probabilities

We compute the *posteriors* for $P(A_j|C3_i, B1_i)$ and report this in Table 8.5.
We validate the *posteriors* for $P(B1_j|C3_i, A_i)$ and report this in Table 8.6.

Table 8.5 Posterior probabilities

$P(A_j\|C3_i, B1_i)$				
A_j				
$C3_i$	$B1_i$	True	False	Totals
True	True	0.092733[a]	0.907267	1.000000
True	False	0.099002	0.900998	1.000000
False	True	0.102723	0.897277	1.000000
False	False	0.109586	0.890414	1.000000

[a]From Table 8.4, $0.092733 = 0.014062/(0.014062 + 0.137579)$

Table 8.6 Posterior probabilities

$P(B1_j\|C3_i, A_i)$				
$B1_j$				
$C3_i$	A_i	True	False	Totals
True	True	0.289345[a]	0.710655	1.000000
True	False	0.304445	0.695555	1.000000
False	True	0.289345	0.710655	1.000000
False	False	0.304445	0.695555	1.000000

[a]From Table 8.4, $0.289345 = 0.014062/(0.014062 + 0.034538)$

8.4.2 Path 2: BBN Solution Protocol

Step 2: Specify the BBN–[$A \rightarrow B1 | B1 \rightarrow C2$]

See Fig. 8.2.

Fig. 8.2 Paths [$A \rightarrow B1 | B1 \rightarrow C2$]

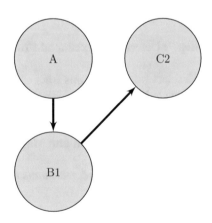

Table 8.7 Node A: prior probabilities

$P(A_j)$		
True	False	Total
0.102300[a]	0.897700	1.000000

[a]From Table 6.5, 0.102300 = 1023/10,000

Table 8.8 Node $B1$: likelihood probabilities

$P(B1_j\|A_i)$			
$B1_j$			
A_i	True	False	Total
True	0.289345[a]	0.710655	1.000000
False	0.304445	0.695555	1.000000

[a]From Table 6.8, 0.289345 = 296/1023

Table 8.9 Node $C2$: likelihood probabilities

$P(C2_j\|B1_i, A_i)$				
$C2_j$				
$B1_i$	A_i	True	False	Total
True	True	0.481677[a]	0.518323	1.000000
True	False	0.481677	0.518323	1.000000
False	True	0.508679	0.491321	1.000000
False	False	0.508679	0.491321	1.000000

[a]From Table 6.15, 0.481677 = 1459/3029

Step 6: Determine Prior or Unconditional Probabilities

We select Node A for this BBN structure as reported in Table 8.7.

Step 7: Determine Likelihood Probabilities

We select Node $B1$ for this BBN structure as reported in Table 8.8.
We select Node $C2$ for this BBN structure as reported in Table 8.9.

Step 8: Determine Joint and Marginal Probabilities

We compute the *joints* and *marginals* for $P(C2_j, B1_i, A_i)$ and report these in Table 8.10.

Step 9: Compute Posterior Probabilities

We compute the *posteriors* for $P(B1_j|C2_i, A_i)$ and report this in Table 8.11.
We compute the *posteriors* for $P(A_j|C2_i, B1_i)$ and report this in Table 8.12.

Table 8.10 Joint and marginal probabilities

$P(C2_j, B1_i, A_i)$				
$C2_j$				
$B1_i$	A_i	True	False	Marginals
True	True	0.014258[a]	0.015342	0.029600
True	False	0.131642	0.141658	0.273300
False	True	0.036981	0.035719	0.072700
False	False	0.317619	0.306781	0.624400
Marginals		0.500500	0.499500	1.000000

Note: ZSMS $= 0.000000 + 0.000000 = 0.000000$
[a]From Tables 8.7, 8.8, and 8.9, $0.014258 = 0.481677 * 0.289345 * 0.102300$

Table 8.11 Posterior probabilities

$P(B1_j|C2_i, A_i)$				
$B1_j$				
$C2_i$	A_i	True	False	Total
True	True	0.278260[a]	0.721740	1.000000
True	False	0.293020	0.706980	1.000000
False	True	0.300469	0.699531	1.000000
False	False	0.315891	0.684109	1.000000

[a]From Table 8.10, $0.278260 = 0.014258/(0.014258 + 0.036981)$

Table 8.12 Posterior probabilities

$P(A_j|C2_i, B1_i)$				
A_j				
$C2_i$	$B1_i$	True	False	Total
True	True	0.097722[a]	0.902278	1.000000
True	False	0.104289	0.895711	1.000000
False	True	0.097722	0.902278	1.000000
False	False	0.104289	0.895711	1.000000

[a]From Table 8.10, $0.097722 = 0.014258/(0.014258 + 0.131642)$

8.4.3   Path 3: BBN Solution Protocol

Step 2: Specify the BBN–$[A \rightarrow C1|B2 \rightarrow C1]$

See Fig. 8.3.

Fig. 8.3 Paths
$[A \rightarrow C1|B2 \rightarrow C1]$

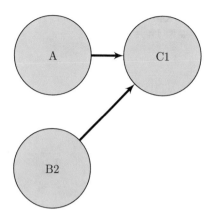

Step 6: Determine Prior or Unconditional Probabilities

We select Node *A* for this BBN structure as reported in Table 8.13.

Table 8.13 Node *A*: prior probabilities

$P(A_j)$		
True	False	Total
0.102300[a]	0.897700	1.000000

[a]From Table 6.5, 0.102300 = 1023/10,000

Table 8.14 Node *B2*: likelihood probabilities

$P(B2_j|A_i)$			
$B2_j$			
A_i	True	False	Total
True	0.302900[a]	0.697100	1.000000
False	0.302900	0.697100	1.000000

[a]From Table 6.10, 0.302900 = 3029/10,000

Table 8.15 Node *C1*: likelihood probabilities

$P(C1_j|B2_i, A_i)$				
$C1_j$				
$B2_i$	A_i	True	False	Total
True	True	0.466216[a]	0.533784	1.000000
True	False	0.483352	0.516648	1.000000
False	True	0.478680	0.521320	1.000000
False	False	0.512172	0.487828	1.000000

[a]From Table 6.13, 0.466216 = 138/296

Step 7: Determine Likelihood Probabilities

We select Node *B2* for this BBN structure as reported in Table 8.14.
We select Node *C1* for this BBN structure as reported in Table 8.15.

Table 8.16 Joint and
marginal probabilities

$P(C1_j, B2_i, A_i)$				
$C1_j$				
$B2_i$	A_i	True	False	Marginals
True	True	0.014446[a]	0.016540	0.030987
True	False	0.131430	0.140484	0.271913
False	True	0.034136	0.037177	0.071313
False	False	0.320510	0.305276	0.625787
Marginals		0.500523	0.499477	1.000000

Note: ZSMS $= -0.000023 + 0.000023 = 0.000000$
[a]From Tables 8.13, 8.14, and 8.15, $0.014446 = 0.466216 * 0.302900 * 0.102300$

Table 8.17 Posterior
probabilities

$P(B2_j | C1_i, A_i)$				
$B2_j$				
$C1_i$	A_i	True	False	Total
True	True	0.297359[a]	0.702641	1.000000
True	False	0.290812	0.709188	1.000000
False	True	0.307912	0.692088	1.000000
False	False	0.315155	0.684845	1.000000

[a]From Table 8.16, $0.297359 = 0.014446/(0.014446 + 0.034136)$

Table 8.18 Posterior
probabilities

$P(A_j | C1_i, B2_i)$				
A_j				
$C1_i$	$B2_i$	True	False	Total
True	True	0.099032[a]	0.900968	1.000000
True	False	0.096254	0.903746	1.000000
False	True	0.105336	0.894664	1.000000
False	False	0.108561	0.891439	1.000000

[a]From Table 8.16, $0.099032 = 0.014446/(0.014446 + 0.131430)$

Step 8: Determine Joint and Marginal Probabilities

We compute the *joints* and *marginals* for $P(C1_j, B2_i, A_i)$ and report these in Table 8.16.
We compute the *posteriors* for $P(B2_j | C1_i, A_i)$ and report this in Table 8.17.
We compute the *posteriors* for $P(A_j | C1_i, B2_i)$ and report this in Table 8.18.

8.5   3-Path BBN

8.5.1   Path 1: BBN Solution Protocol

Step 2: Specify the BBN–$[A \rightarrow B1|A \rightarrow C1|B1 \rightarrow C1]$

See Fig. 8.4.

Fig. 8.4 Paths $[A \rightarrow B1|A \rightarrow C1|B1 \rightarrow C1]$

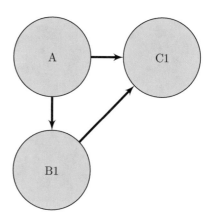

Step 6: Determine Prior or Unconditional Probabilities

We select Node A for this BBN structure as reported in Table 8.19.

Step 7: Determine Likelihood Probabilities

We select Node $B1$ for this BBN structure as reported in Table 8.20.
We select Node $C1$ for this BBN structure as reported in Table 8.21.

Step 8: Determine Joint and Marginal Probabilities

We compute the *joints* and *marginals* for $P(C1_j, B1_i, A_i)$ and report these in Table 8.22.

Table 8.19 Node A: prior probabilities

$P(A_j)$		
True	False	Total
0.102300[a]	0.897700	1.000000

[a]From Table 6.5, $0.102300 = 1023/10{,}000$

Table 8.20 Node $B1$: likelihood probabilities

$P(B1_j\|A_i)$			
$B1_j$			
A_i	True	False	Total
True	0.289345[a]	0.710655	1.000000
False	0.304445	0.695555	1.000000

[a]From Table 6.8, $0.289345 = 296/1023$

Table 8.21 Node $C1$: likelihood probabilities

$P(C1_j\|B1_i,A_i)$				
$C1_j$				
$B1_i$	A_i	True	False	Total
True	True	0.466216[a]	0.533784	1.000000
True	False	0.483352	0.516648	1.000000
False	True	0.478680	0.521320	1.000000
False	False	0.512172	0.487828	1.000000

[a]From Table 6.13, $0.466216 = 138/296$

Table 8.22 Joint and marginal probabilities

$P(C1_j,B1_i,A_i)$				
$C1_j$				
$B1_i$	A_i	True	False	Marginals
True	True	0.013800[a]	0.015800	0.029600
True	False	0.132100	0.141200	0.273300
False	True	0.034800	0.037900	0.072700
False	False	0.319800	0.304600	0.624400
Marginals		0.500500	0.499500	1.000000

Note: ZSMS $= 0.000000 + 0.000000 = 0.000000$
[a]From Tables 8.19, 8.20, and 8.21, $0.013800 = 0.466216 * 0.289345 * 0.102300$

Table 8.23 Posterior probabilities

$P(B1_j\|C1_i,A_i)$				
$B1_j$				
$C1_i$	A_i	True	False	Total
True	True	0.283951[a]	0.716049	1.000000
True	False	0.292321	0.707679	1.000000
False	True	0.294227	0.705773	1.000000
False	False	0.316734	0.683266	1.000000

[a]From Table 8.22, $0.283951 = 0.013800/(0.013800 +0.034800)$

Step 9: Compute Posterior Probabilities

We compute the *posteriors* for $P(B1_j|C1_i,A_i)$ and report this in Table 8.23.
We compute the *posteriors* for $P(A_j|C1_i,B1_i)$ and report this in Table 8.24.

Table 8.24 Posterior
probabilities

$P(A_j|C1_i, B1_i)$				
A_j				
$C1_i$	$B1_i$	True	False	Total
True	True	0.094585[a]	0.905415	1.000000
True	False	0.098139	0.901861	1.000000
False	True	0.100637	0.899363	1.000000
False	False	0.110657	0.889343	1.000000

[a]From Table 8.22, $0.094585 = 0.013800/(0.013800 +0.132100)$

Chapter 9
4-Node BBN

9.1 Introduction

This chapter represents specification and presentation on a 4-Node Bayesian Belief Networks using the BBN solution protocol with counts and likelihood probabilities we created from the synthetic node process developed in Chap. 6.

9.2 Proof

$$P(A|B, C, D) = \frac{P(D, C, B, A)}{P(B, C, D)} \tag{9.1}$$

$$PA, B, C, D = P(D, C, B, A) \tag{9.2}$$

$$PA, B, C, D = P(A|B, C, D)P(B|C, D)P(C|D)P(D) \tag{9.3}$$

$$P(D, C, B, A) = P(D|C, B, A)P(C|B, A)P(B|A)P(A) \tag{9.4}$$

$$P(A|B, C, D)P(B|C, D)P(C|D)P(D) = P(D|C, B, A)P(C|B, A)P(B|A)P(A) \tag{9.5}$$

$$P(P(A|B, C, D) = \frac{P(D|C, B, A)P(C|B, A)P(B|A)P(A)}{P(B|C, D)P(C|D)P(D)} \tag{9.6}$$

$$P(P(A|B, C, D) = \frac{P(D|C, B, A)P(C|B, A)P(B, A)}{P(B|C, D)P(C|D)P(D)} \tag{9.7}$$

$$P(P(A|B, C, D) = \frac{P(D|C, B, A)P(C, B, A)}{P(B|C, D)P(C|D)P(D)} \tag{9.8}$$

$$P(P(A|B, C, D) = \frac{P(D, C, B, A)}{P(B|C, D)P(C|D)P(D)} \tag{9.9}$$

© Springer International Publishing AG 2016
J. Grover, *The Manual of Strategic Economic Decision Making*,
DOI 10.1007/978-3-319-48414-3_9

$$P(P(A|B,C,D) = \frac{P(D,C,B,A)}{P(B|C,D)P(C,D)} \qquad (9.10)$$

$$P(P(A|B,C,D) = \frac{P(D,C,B,A)}{P(B,C,D)} \Box \qquad (9.11)$$

9.3   BBN Solution Protocol

- Step 2: Specify the BBN
- Step 6: Determine prior or *unconditional* probabilities
- Step 7: Determine likelihood probabilities
- Step 8: Determine joint and marginal probabilities
- Step 9: Compute posterior probabilities

9.4   3-Path BBN

9.4.1   Path 1: BBN Solution Protocol

Step 2: Specify the BBN–[$A \rightarrow B1|A \rightarrow C3|A \rightarrow D7$]

See Fig. 9.1.

Step 6: Determine Prior or Unconditional Probabilities

We select Node *A* for this BBN structure as reported in Table 9.1.

Fig. 9.1 Paths
[$A \rightarrow B1|A \rightarrow C3|A \rightarrow D7$]

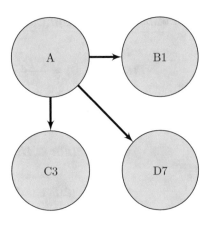

Table 9.1 Node A: prior
probabilities

$P(A_j)$		
True	False	Total
0.102300[a]	0.897700	1.000000

[a]From Table 6.5, 0.102300 = 1023/10000

Step 7: Determine Likelihood Probabilities

We select Node $B1$ for this BBN structure as reported in Table 9.2.

Table 9.2 Node $B1$:
likelihood probabilities

$P(B1_j|A_i)$			
$B1_j$			
A_i	True	False	Total
True	0.289345[a]	0.710655	1.000000
False	0.304445	0.695555	1.000000

[a]From Table 6.8, 0.289345 = 296/1023

We select Node $C3$ for this BBN structure as reported in Table 9.3.

Table 9.3 Node $C3$:
likelihood probabilities

$P(C3_j|\cancel{B1_i}, A_i)$				
$C3_j$				
$\cancel{B1_i}$	A_i	True	False	Total
~~True~~	True	0.475073[a]	0.524927	1.000000
~~True~~	False	0.503398	0.496602	1.000000
~~False~~	~~True~~	~~0.475073~~	~~0.524927~~	~~1.000000~~
~~False~~	~~False~~	~~0.503398~~	~~0.496602~~	~~1.000000~~

[a]From Table 6.17, 0.475073 = 486/1023

We select Node $D7$ for this BBN structure as reported in Table 9.4.

Step 8: Determine Joint and Marginal Probabilities

We compute the *joints* and *marginals* for $P(D7_j, C3_i, B1_i, A_i)$ and report these in
Table 9.5.

Step 9: Compute Posterior Probabilities

We compute the *posteriors* for $P(C3_j|D7_i, B1_i, A_i)$ and report this in Table 9.6.

Table 9.4 Node $D7$:
likelihood probabilities

$P(D7_j | C3_i, B1_i, A_i)$

$D7_j$

$C3_i$	$B1_i$	A_i	True	False	Total
True	True	True	0.730205[a]	0.269795	1.000000
True	True	False	0.711485	0.288515	1.000000
True	False	True	0.730205	0.269795	1.000000
True	False	False	0.711485	0.288515	1.000000
False	True	True	0.730205	0.269795	1.000000
False	True	False	0.711485	0.288515	1.000000
False	False	True	0.730205	0.269795	1.000000
False	False	False	0.711485	0.288515	1.000000

[a]From Table 6.34, 0.730205 = 747/1023

Table 9.5 Joint and marginal
probabilities

$P(D7_j, C3_i, B1_i, A_i)$

$D7_j$

$C3_i$	$B1_i$	A_i	True	False	Marginals
True	True	True	0.010268[a]	0.003794	0.014062
True	True	False	0.097885	0.039693	0.137579
True	False	True	0.025220	0.009318	0.034538
True	False	False	0.223635	0.090686	0.314321
False	True	True	0.011346	0.004192	0.015538
False	True	False	0.096564	0.039158	0.135721
False	False	True	0.027866	0.010296	0.038162
False	False	False	0.220616	0.089462	0.310079
Marginals			0.713400	0.286600	1.000000

Note: ZSMS = 0.000000 + 0.000000 = 0.000000
[a]From Tables 9.1, 9.2, 9.3, and 9.4, 0.010268 = 0.730205 *
0.475073 * 0.289345 * 0.102300

Table 9.6 Posterior
probabilities

$P(C3_j | D7_i, B1_i, A_i)$

$C3_j$

$D7_i$	$B1_i$	A_i	True	False	Total
True	True	True	0.475073[a]	0.524927	1.000000
True	True	False	0.503398	0.496602	1.000000
True	False	True	0.475073	0.524927	1.000000
True	False	False	0.503398	0.496602	1.000000
False	True	True	0.475073	0.524927	1.000000
False	True	False	0.503398	0.496602	1.000000
False	False	True	0.475073	0.524927	1.000000
False	False	False	0.503398	0.496602	1.000000

[a]From Table 9.5, 0.475073 = 0.010268/(0.010268 +
0.011346)

We compute the *posteriors* for $P(B1_j|D7_i, C3_i, A_i)$ and report this in Table 9.7.

Table 9.7 Posterior
probabilities

$P(B1_j|D7_i, C3_i, A_i)$					
$B1_j$					
$D7_i$	$C3_i$	A_i	True	False	Total
True	True	True	0.289345[a]	0.710655	1.000000
True	True	False	0.304445	0.695555	1.000000
True	False	True	0.289345	0.710655	1.000000
True	False	False	0.304445	0.695555	1.000000
False	True	True	0.289345	0.710655	1.000000
False	True	False	0.304445	0.695555	1.000000
False	False	True	0.289345	0.710655	1.000000
False	False	False	0.304445	0.695555	1.000000

[a]From Table 9.5, 0.289345 = 0.010268/(0.010268 + 0.025220)

We compute the *posteriors* for $P(A_j|D7_i, C3_i, B1_i)$ and report this in Table 9.8.

Table 9.8 Posterior
probabilities

$P(A_j|D7_i, C3_i, B1_i)$					
A_j					
$D7_i$	$C3_i$	$B1_i$	True	False	Total
True	True	True	0.094942[a]	0.905058	1.000000
True	True	False	0.101343	0.898657	1.000000
True	False	True	0.105142	0.894858	1.000000
True	False	False	0.112146	0.887854	1.000000
False	True	True	0.087241	0.912759	1.000000
False	True	False	0.093177	0.906823	1.000000
False	False	True	0.096702	0.903298	1.000000
False	False	False	0.103209	0.896791	1.000000

[a]From Table 9.5, 0.094942 = 0.010268/(0.010268 + 0.097885)

9.4.2   Path 3: BBN Solution Protocol

Step 2: Specify the BBN–[$A \rightarrow B1|A \rightarrow C3|B1 \rightarrow D6$]

See Fig. 9.2.

Fig. 9.2 Paths $[A \rightarrow B1|A \rightarrow C3|B1 \rightarrow D6]$

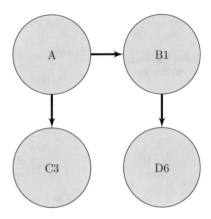

Step 6: Determine Prior or Unconditional Probabilities

We select Node *A* for this BBN structure as reported in Table 9.9.

Table 9.9 Node *A*: prior probabilities

$P(A_j)$		
True	False	Total
0.102300[a]	0.897700	1.000000

[a]From Table 6.5, 0.102300 = 1023/10000

Step 7: Determine Likelihood Probabilities

We select Node *B*1 for this BBN structure as reported in Table 9.10.

Table 9.10 Node *B*1: likelihood probabilities

$P(B1_j|A_i)$			
$B1_j$			
A_i	True	False	Total
True	0.289345[a]	0.710655	1.000000
False	0.304445	0.695555	1.000000

[a]From Table 6.8, 0.289345 = 296/1023

We select Node $C3$ for this BBN structure as reported in Table 9.11.

Table 9.11 Node $C3$:
likelihood probabilities

$P(C3_j \| BX_i, A_i)$				
$C3_j$				
BX_i	A_i	True	False	Total
~~True~~	True	0.475073[a]	0.524927	1.000000
~~True~~	False	0.503398	0.496602	1.000000
~~False~~	~~True~~	~~0.475073~~	~~0.524927~~	~~1.000000~~
~~False~~	~~False~~	~~0.503398~~	~~0.496602~~	~~1.000000~~

[a]From Table 6.17, 0.475073 = 486/1023

We select Node $D6$ for this BBN structure as reported in Table 9.12.

Table 9.12 Node $D6$:
likelihood probabilities

$P(D6_j \| C3_i, B1_i, A_i)$					
$D6_j$					
$C3_i$	$B1_i$	A_i	True	False	Total
~~True~~	True	~~True~~	0.704853[a]	0.295147	1.000000
~~True~~	~~True~~	~~False~~	~~0.704853~~	~~0.295147~~	~~1.000000~~
~~True~~	False	~~True~~	0.717114	0.282886	1.000000
~~True~~	~~False~~	~~False~~	~~0.717114~~	~~0.282886~~	~~1.000000~~
~~False~~	~~True~~	~~True~~	~~0.704853~~	~~0.295147~~	~~1.000000~~
~~False~~	~~True~~	~~False~~	~~0.704853~~	~~0.295147~~	~~1.000000~~
~~False~~	~~False~~	~~True~~	~~0.717114~~	~~0.282886~~	~~1.000000~~
~~False~~	~~False~~	~~False~~	~~0.717114~~	~~0.282886~~	~~1.000000~~

[a]From Table 6.32, 0.704853 = 2135/3029

Step 8: Determine Joint and Marginal Probabilities

We compute the *joints* and *marginals* for $P(D6_j, C3_i, B1_i, A_i)$ and report these in
Table 9.13.

Table 9.13 Joint and marginal probabilities

$P(D6_j, C3_i, B1_i, A_i)$					
$D6_j$					
$C3_i$	$B1_i$	A_i	True	False	Marginals
True	True	True	0.009912[a]	0.004150	0.014062
True	True	False	0.096973	0.040606	0.137579
True	False	True	0.024768	0.009770	0.034538
True	False	False	0.225404	0.088917	0.314321
False	True	True	0.010952	0.004586	0.015538
False	True	False	0.095664	0.040058	0.135721
False	False	True	0.027367	0.010796	0.038162
False	False	False	0.222362	0.087717	0.310079
Marginals			0.713400	0.286600	1.000000

Note: ZSMS = 0.000000 + 0.000000 = 0.000000
[a]From Tables 9.9, 9.10, 9.11, and 9.12, 0.009912 = 0.704853 * 0.475073 * 0.289345 * 0.102300

Step 9: Compute Posterior Probabilities

We compute the *posteriors* for $P(C3_j | D6_i, B1_i, A_i)$ and report this in Table 9.14.

Table 9.14 Posterior probabilities

$P(C3_j | D6_i, B1_i, A_i)$					
$C3_j$					
$D6_i$	$B1_i$	A_i	True	False	Total
True	True	True	0.475073[a]	0.524927	1.000000
True	True	False	0.503398	0.496602	1.000000
True	False	True	0.475073	0.524927	1.000000
True	False	False	0.503398	0.496602	1.000000
False	True	True	0.475073	0.524927	1.000000
False	True	False	0.503398	0.496602	1.000000
False	False	True	0.475073	0.524927	1.000000
False	False	False	0.503398	0.496602	1.000000

[a]From Table 9.13, 0.475073 = 0.009912/(0.009912 + 0.010952)

We compute the *posteriors* for $P(B1_j | D6_i, C3_i, A_i)$ and report this in Table 9.15.

Table 9.15 Posterior probabilities

$P(B1_j$|$D6_i, C3_i, A_i)$					
$B1_j$					
$D6_i$	$C3_i$	A_i	True	False	Total
True	True	True	0.285812[a]	0.714188	1.000000
True	True	False	0.300805	0.699195	1.000000
True	False	True	0.285812	0.714188	1.000000
True	False	False	0.300805	0.699195	1.000000
False	True	True	0.298147	0.701853	1.000000
False	True	False	0.313503	0.686497	1.000000
False	False	True	0.298147	0.701853	1.000000
False	False	False	0.313503	0.686497	1.000000

[a]From Table 9.13, 0.285812 = 0.009912/(0.009912 + 0.024768)

We compute the *posteriors* for $P(A_j|D6_i, C3_i, B1_i)$ and report this in Table 9.16.

Table 9.16 Posterior probabilities

$P(A_j$|$D6_i, C3_i, B1_i)$					
A_j					
$D6_i$	$C3_i$	$B1_i$	True	False	Total
True	True	True	0.092733[a]	0.907267	1.000000
True	True	False	0.099002	0.900998	1.000000
True	False	True	0.102723	0.897277	1.000000
True	False	False	0.109586	0.890414	1.000000
False	True	True	0.092733	0.907267	1.000000
False	True	False	0.099002	0.900998	1.000000
False	False	True	0.102723	0.897277	1.000000
False	False	False	0.109586	0.890414	1.000000

[a]From Table 9.13, 0.092733 = 0.009912/(0.009912 + 0.096973)

9.4.3 Path 4: BBN Solution Protocol

Step 2: Specify the BBN–[$A \rightarrow B1|A \rightarrow C3|C3 \rightarrow D5$]

See Fig. 9.3.

Fig. 9.3 Paths $[A \rightarrow$
$B1|A \rightarrow C3|C3 \rightarrow D5]$

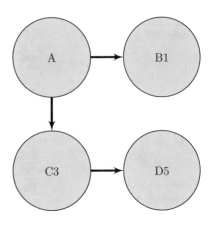

Step 6: Determine Prior or Unconditional Probabilities

We select Node A for this BBN structure as reported in Table 9.17.

Table 9.17 Node A: prior
probabilities

$P(A_j)$		
True	False	Total
0.102300[a]	0.897700	1.000000

[a]From Table   6.5,   0.102300   =
1023/10000

Step 7: Determine Likelihood Probabilities

We select Node $B1$ for this BBN structure as reported in Table 9.18.

Table 9.18 Node $B1$:
likelihood probabilities

$P(B1_j|A_i)$			
$B1_j$			
A_i	True	False	Total
True	0.289345[a]	0.710655	1.000000
False	0.304445	0.695555	1.000000

[a]From Table 6.8, 0.289345 = 296/1023

We select Node $C3$ for this BBN structure as reported in Table 9.19.

Table 9.19 Node $C3$: likelihood probabilities

$P(C3_j\|B1_i, A_i)$				
$C3_j$				
$B1_i$	A_i	True	False	Total
True	True	0.475073[a]	0.524927	1.000000
True	False	0.503398	0.496602	1.000000
False	True	0.475073	0.524927	1.000000
False	False	0.503398	0.496602	1.000000

[a]From Table 6.17, 0.475073 = 486/1023

We select Node $D5$ for this BBN structure as reported in Table 9.20.

Table 9.20 Node $D5$: likelihood probabilities

$P(D5_j\|C3_i, B1_i, A_i)$					
$D5_j$					
$C3_i$	$B1_i$	A_i	True	False	Total
True	True	True	0.710490[a]	0.289510	1.000000
True	True	False	0.710490	0.289510	1.000000
True	False	True	0.710490	0.289510	1.000000
True	False	False	0.710490	0.289510	1.000000
False	True	True	0.716316	0.283684	1.000000
False	True	False	0.716316	0.283684	1.000000
False	False	True	0.716316	0.283684	1.000000
False	False	False	0.716316	0.283684	1.000000

[a]From Table 6.30, 0.710490 = 3556/5005

Step 8: Determine Joint and Marginal Probabilities

We compute the *joints* and *marginals* for $P(D5_j, C3_i, B1_i, A_i)$ and report these in Table 9.21.

Table 9.21 Joint and marginal probabilities

			$P(D5_j, C3_i, B1_i, A_i)$		
			$D5_j$		
$C3_i$	$B1_i$	A_i	True	False	Marginals
True	True	True	0.009991[a]	0.004071	0.014062
True	True	False	0.097748	0.039830	0.137579
True	False	True	0.024539	0.009999	0.034538
True	False	False	0.223322	0.090999	0.314321
False	True	True	0.011130	0.004408	0.015538
False	True	False	0.097219	0.038502	0.135721
False	False	True	0.027336	0.010826	0.038162
False	False	False	0.222114	0.087964	0.310079
Marginals			0.713400	0.286600	1.000000

Note: ZSMS = 0.000000 + 0.000000 = 0.000000
[a]From Tables 9.17, 9.18, 9.19, and 9.20, 0.009991 = 0.710490 * 0.475073 * 0.289345 * 0.102300

Step 9: Compute Posterior Probabilities

We compute the *posteriors* for $P(C3_j|D5_i, B1_i, A_i)$ and report this in Table 9.22.

Table 9.22 Posterior probabilities

			$P(C3_j|D5_i, B1_i, A_i)$		
			$C3_j$		
$D5_i$	$B1_i$	A_i	True	False	Total
True	True	True	0.473037[a]	0.526963	1.000000
True	True	False	0.501356	0.498644	1.000000
True	False	True	0.473037	0.526963	1.000000
True	False	False	0.501356	0.498644	1.000000
False	True	True	0.480146	0.519854	1.000000
False	True	False	0.508480	0.491520	1.000000
False	False	True	0.480146	0.519854	1.000000
False	False	False	0.508480	0.491520	1.000000

[a]From Table 9.21, 0.473037 = 0.009991/(0.009991 + 0.011130)

We compute the *posteriors* for $P(B1_j|D5_i, C3_i, A_i)$ and report this in Table 9.23.

Table 9.23 Posterior probabilities

$P(B1_j|D5_i, C3_i, A_i)$

			$B1_j$		
$D5_i$	$C3_i$	A_i	True	False	Total
True	True	True	0.289345[a]	0.710655	1.000000
True	True	False	0.304445	0.695555	1.000000
True	False	True	0.289345	0.710655	1.000000
True	False	False	0.304445	0.695555	1.000000
False	True	True	0.289345	0.710655	1.000000
False	True	False	0.304445	0.695555	1.000000
False	False	True	0.289345	0.710655	1.000000
False	False	False	0.304445	0.695555	1.000000

[a]From Table 9.21, 0.289345 = 0.009991/(0.009991 + 0.024539)

We compute the *posteriors* for $P(A_j|D5_i, C3_i, B1_i)$ and report this in Table 9.24.

Table 9.24 Posterior probabilities

$P(A_j|D5_i, C3_i, B1_i)$

			A_j		
$D5_i$	$C3_i$	$B1_i$	True	False	Total
True	True	True	0.092733[a]	0.907267	1.000000
True	True	False	0.099002	0.900998	1.000000
True	False	True	0.102723	0.897277	1.000000
True	False	False	0.109586	0.890414	1.000000
False	True	True	0.092733	0.907267	1.000000
False	True	False	0.099002	0.900998	1.000000
False	False	True	0.102723	0.897277	1.000000
False	False	False	0.109586	0.890414	1.000000

[a]From Table 9.21, 0.092733 = 0.009991/(0.009991 + 0.097748)

9.4.4   Path 5: BBN Solution Protocol

Step 2: Specify the BBN–[$A \rightarrow B1|A \rightarrow D7|B1 \rightarrow C2$]

See Fig. 9.4.

Fig. 9.4 Paths $[A \rightarrow$
$B1|A \rightarrow D7|B1 \rightarrow C2]$

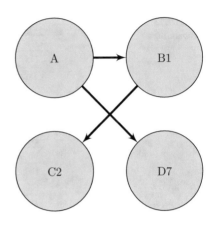

Step 6: Determine Prior or Unconditional Probabilities

We select Node A for this BBN structure as reported in Table 9.25.

Table 9.25 Node A: prior probabilities

$P(A_j)$		
True	False	Total
0.102300[a]	0.897700	1.000000

[a]From Table 6.5, 0.102300 = 1023/10000

Step 7: Determine Likelihood Probabilities

We select Node $B1$ for this BBN structure as reported in Table 9.26.

Table 9.26 Node $B1$: likelihood probabilities

$P(B1_j|A_i)$			
$B1_j$			
A_i	True	False	Total
True	0.289345[a]	0.710655	1.000000
False	0.304445	0.695555	1.000000

[a]From Table 6.8, 0.289345 = 296/1023

We select Node $C2$ for this BBN structure as reported in Table 9.27.

Table 9.27 Node $C2$: likelihood probabilities

$P(C2_j|B1_i, A_i)$				
$C2_j$				
$B1_i$	A_i	True	False	Total
True	True	0.481677[a]	0.518323	1.000000
True	False	0.481677	0.518323	1.000000
False	True	0.508679	0.491321	1.000000
False	False	0.508679	0.491321	1.000000

[a]From Table 6.15, 0.481677 = 1459/3029

We select Node $D7$ for this BBN structure as reported in Table 9.28.

Table 9.28 Node $D7$: likelihood probabilities

$P(D7_j|C2_i, B1_i, A_i)$					
$D7_j$					
$C2_i$	$B1_i$	A_i	True	False	Total
True	True	True	0.730205[a]	0.269795	1.000000
True	True	False	0.711485	0.288515	1.000000
True	False	True	0.730205	0.269795	1.000000
True	False	False	0.711485	0.288515	1.000000
False	True	True	0.730205	0.269795	1.000000
False	True	False	0.711485	0.288515	1.000000
False	False	True	0.730205	0.269795	1.000000
False	False	False	0.711485	0.288515	1.000000

[a]From Table 6.34, 0.730205 = 747/1023

Step 8: Determine Joint and Marginal Probabilities

We compute the *joints* and *marginals* for $P(D7_j, C2_i, B1_i, A_i)$ and report these in Table 9.29.

Table 9.29 Joint and marginal probabilities

$P(D7_j, C2_i, B1_i, A_i)$					
$D7_j$					
$C2_i$	$B1_i$	A_i	True	False	Marginals
True	True	True	0.010411[a]	0.003847	0.014258
True	True	False	0.093662	0.037981	0.131642
True	False	True	0.027004	0.009977	0.036981
True	False	False	0.225981	0.091638	0.317619
False	True	True	0.011203	0.004139	0.015342
False	True	False	0.100787	0.040870	0.141658
False	False	True	0.026082	0.009637	0.035719
False	False	False	0.218270	0.088511	0.306781
Marginals			0.713400	0.286600	1.000000

Note: ZSMS = 0.000000 + 0.000000 = 0.000000
[a]From Tables 9.25, 9.26, 9.27, and 9.28, 0.010411 = 0.730205
* 0.481677 * 0.289345 * 0.102300

Step 9: Compute Posterior Probabilities

We compute the *posteriors* for $P(C2_j|D7_i, B1_i, A_i)$ and report this in Table 9.30.

Table 9.30 Posterior probabilities

$P(C2_j|D7_i, B1_i, A_i)$					
$C2_j$					
$D7_i$	$B1_i$	A_i	True	False	Total
True	True	True	0.481677[a]	0.518323	1.000000
True	True	False	0.481677	0.518323	1.000000
True	False	True	0.508679	0.491321	1.000000
True	False	False	0.508679	0.491321	1.000000
False	True	True	0.481677	0.518323	1.000000
False	True	False	0.481677	0.518323	1.000000
False	False	True	0.508679	0.491321	1.000000
False	False	False	0.508679	0.491321	1.000000

[a]From Table 9.29, 0.481677 = 0.010411/(0.010411 + 0.011203)

We compute the *posteriors* for $P(B1_j|D7_i, C2_i, A_i)$ and report this in Table 9.31.

Table 9.31 Posterior probabilities

$P(B1_j|D7_i, C2_i, A_i)$

			$B1_j$		
$D7_i$	$C2_i$	A_i	True	False	Total
True	True	True	0.278260[a]	0.721740	1.000000
True	True	False	0.293020	0.706980	1.000000
True	False	True	0.300469	0.699531	1.000000
True	False	False	0.315891	0.684109	1.000000
False	True	True	0.278260	0.721740	1.000000
False	True	False	0.293020	0.706980	1.000000
False	False	True	0.300469	0.699531	1.000000
False	False	False	0.315891	0.684109	1.000000

[a]From Table 9.29, $0.278260 = 0.010411/(0.010411 + 0.027004)$

We compute the *posteriors* for $P(A_j|D7_i, C2_i, B1_i)$ and report this in Table 9.32.

Table 9.32 Posterior probabilities

$P(A_j|D7_i, C2_i, B1_i)$

			A_j		
$D7_i$	$C2_i$	$B1_i$	True	False	Total
True	True	True	0.100036[a]	0.899964	1.000000
True	True	False	0.106740	0.893260	1.000000
True	False	True	0.100036	0.899964	1.000000
True	False	False	0.106740	0.893260	1.000000
False	True	True	0.091964	0.908036	1.000000
False	True	False	0.098187	0.901813	1.000000
False	False	True	0.091964	0.908036	1.000000
False	False	False	0.098187	0.901813	1.000000

[a]From Table 9.29, $0.100036 = 0.010411/(0.010411 + 0.093662)$

9.4.5 Path 7: BBN Solution Protocol

Step 2: Specify the BBN–[$A \rightarrow B1|A \rightarrow D3|C4 \rightarrow D3$]

See Fig. 9.5.

Fig. 9.5 Paths [$A \rightarrow$ $B1|A \rightarrow D3|C4 \rightarrow D3$]

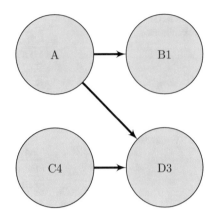

Step 6: Determine Prior or Unconditional Probabilities

We select Node *A* for this BBN structure as reported in Table 9.33.

Table 9.33 Node *A*: prior probabilities

$P(A_j)$		
True	False	Total
0.102300[a]	0.897700	1.000000

[a]From Table 6.5, 0.102300 = 1023/10000

Step 7: Determine Likelihood Probabilities

We select Node *B*1 for this BBN structure as reported in Table 9.34.

Table 9.34 Node *B*1: likelihood probabilities

$P(B1_j|A_i)$			
$B1_j$			
A_i	True	False	Total
True	0.289345[a]	0.710655	1.000000
False	0.304445	0.695555	1.000000

[a]From Table 6.8, 0.289345 = 296/1023

We select Node *C*4 for this BBN structure as reported in Table 9.35.

Table 9.35 Node $C4$:
likelihood probabilities

$P(C4_j\|B1_i, A_i)$				
$C4_j$				
$B1_i$	A_i	True	False	Total
True	True	0.500500[a]	0.499500	1.000000
True	False	0.500500	0.499500	1.000000
False	True	0.500500	0.499500	1.000000
False	False	0.500500	0.499500	1.000000

[a]From Table 6.19, $0.500500 = 5005/10000$

We select Node $D3$ for this BBN structure as reported in Table 9.36.

Table 9.36 Node $D3$:
likelihood probabilities

$P(D3_j\|C4_i, B1_i, A_i)$					
$D3_j$					
$C4_i$	$B1_i$	A_i	True	False	Total
True	True	True	0.716049[a]	0.283951	1.000000
True	True	False	0.709892	0.290108	1.000000
True	False	True	0.716049	0.283951	1.000000
True	False	False	0.709892	0.290108	1.000000
False	True	True	0.743017	0.256983	1.000000
False	True	False	0.713100	0.286900	1.000000
False	False	True	0.743017	0.256983	1.000000
False	False	False	0.713100	0.286900	1.000000

[a]From Table 6.26, $0.716049 = 348/486$

Step 8: Determine Joint and Marginal Probabilities

We compute the *joints* and *marginals* for $P(D3_j, C4_i, B1_i, A_i)$ and report these in Table 9.37.

Table 9.37 Joint and marginal probabilities

$P(D3_j, C4_i, B1_i, A_i)$					
$D3_j$					
$C4_i$	$B1_i$	A_i	True	False	Marginals
True	True	True	0.010608[a]	0.004207	0.014815
True	True	False	0.097104	0.039683	0.136787
True	False	True	0.026054	0.010332	0.036386
True	False	False	0.221850	0.090662	0.312512
False	True	True	0.010986	0.003800	0.014785
False	True	False	0.097348	0.039166	0.136513
False	False	True	0.026982	0.009332	0.036314
False	False	False	0.222407	0.089481	0.311888
Marginals			0.713338	0.286662	1.000000

Note: ZSMS = 0.000062 − 0.000062 = 0.000000
[a]From Tables 9.33, 9.34, 9.35, and 9.36, 0.010608 = 0.716049 * 0.500500 * 0.289345 * 0.102300

Step 9: Compute Posterior Probabilities

We compute the *posteriors* for $P(C4_j|D3_i, B1_i, A_i)$ and report this in Table 9.38.

Table 9.38 Posterior probabilities

$P(C4_j|D3_i, B1_i, A_i)$					
$C4_j$					
$D3_i$	$B1_i$	A_i	True	False	Total
True	True	True	0.491259[a]	0.508741	1.000000
True	True	False	0.499373	0.500627	1.000000
True	False	True	0.491259	0.508741	1.000000
True	False	False	0.499373	0.500627	1.000000
False	True	True	0.525425	0.474575	1.000000
False	True	False	0.503280	0.496720	1.000000
False	False	True	0.525425	0.474575	1.000000
False	False	False	0.503280	0.496720	1.000000

[a]From Table 9.37, 0.491259 = 0.010608/(0.010608 + 0.010986)

We compute the *posteriors* for $P(B1_j|D3_i, C4_i, A_i)$ and report this in Table 9.39.

Table 9.39 Posterior probabilities

$P(B1_j|D3_i, C4_i, A_i)$

			$B1_j$		
$D3_i$	$C4_i$	A_i	True	False	Total
True	True	True	0.289345[a]	0.710655	1.000000
True	True	False	0.304445	0.695555	1.000000
True	False	True	0.289345	0.710655	1.000000
True	False	False	0.304445	0.695555	1.000000
False	True	True	0.289345	0.710655	1.000000
False	True	False	0.304445	0.695555	1.000000
False	False	True	0.289345	0.710655	1.000000
False	False	False	0.304445	0.695555	1.000000

[a]From Table 9.37, 0.289345 = 0.010608/(0.010608 + 0.026054)

We compute the *posteriors* for $P(A_j|D3_i, C4_i, B1_i)$ and report this in Table 9.40.

Table 9.40 Posterior probabilities

$P(A_j|D3_i, C4_i, B1_i)$

			A_j		
$D3_i$	$C4_i$	$B1_i$	True	False	Total
True	True	True	0.098486[a]	0.901514	1.000000
True	True	False	0.105099	0.894901	1.000000
True	False	True	0.101406	0.898594	1.000000
True	False	False	0.108191	0.891809	1.000000
False	True	True	0.095847	0.904153	1.000000
False	True	False	0.102302	0.897698	1.000000
False	False	True	0.088433	0.911567	1.000000
False	False	False	0.094441	0.905559	1.000000

[a]From Table 9.37, 0.098486 = 0.010608/(0.010608 + 0.097104)

9.4.6 Path 8: BBN Solution Protocol

Step 2: Specify the BBN–[$A \rightarrow B1|B1 \rightarrow C2|B1 \rightarrow D6$]

See Fig. 9.6.

Fig. 9.6 Paths [$A \rightarrow$
$B1|B1 \rightarrow C2|B1 \rightarrow D6$]

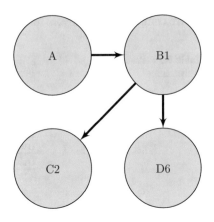

Step 6: Determine Prior or Unconditional Probabilities

We select Node A for this BBN structure as reported in Table 9.41.

Table 9.41 Node A: prior
probabilities

$P(A_j)$		
True	False	Total
0.102300[a]	0.897700	1.000000

[a]From   Table   6.5,   0.102300   =
1023/10000

Step 7: Determine Likelihood Probabilities

We select Node $B1$ for this BBN structure as reported in Table 9.42.

Table 9.42 Node $B1$:
likelihood probabilities

$P(B1_j|A_i)$			
$B1_j$			
A_i	True	False	Total
True	0.289345[a]	0.710655	1.000000
False	0.304445	0.695555	1.000000

[a]From Table 6.8, 0.289345 = 296/1023

We select Node $C2$ for this BBN structure as reported in Table 9.43.

Table 9.43 Node $C2$: likelihood probabilities

$P(C2_j\|B1_i, A_i)$				
$C2_j$				
$B1_i$	A_i	True	False	Total
True	True	0.481677[a]	0.518323	1.000000
True	False	0.481677	0.518323	1.000000
False	True	0.508679	0.491321	1.000000
False	False	0.508679	0.491321	1.000000

[a]From Table 6.15, $0.481677 = 1459/3029$

We select Node $D6$ for this BBN structure as reported in Table 9.44.

Table 9.44 Node $D6$: likelihood probabilities

$P(D6_j\|C2_i, B1_i, A_i)$					
$D6_j$					
$C2_i$	$B1_i$	A_i	True	False	Total
True	True	True	0.704853[a]	0.295147	1.000000
True	True	False	0.704853	0.295147	1.000000
True	False	True	0.717114	0.282886	1.000000
True	False	False	0.717114	0.282886	1.000000
False	True	True	0.704853	0.295147	1.000000
False	True	False	0.704853	0.295147	1.000000
False	False	True	0.717114	0.282886	1.000000
False	False	False	0.717114	0.282886	1.000000

[a]From Table 6.32, $0.704853 = 2135/3029$

Step 8: Determine Joint and Marginal Probabilities

We compute the *joints* and *marginals* for $P(D6_j, C2_i, B1_i, A_i)$ and report these in Table 9.45.

Table 9.45 Joint and
marginal probabilities

$P(D6_j, C2_i, B1_i, A_i)$					
$D6_j$					
$C2_i$	$B1_i$	A_i	True	False	Marginals
True	True	True	0.010050[a]	0.004208	0.014258
True	True	False	0.092789	0.038854	0.131642
True	False	True	0.026520	0.010461	0.036981
True	False	False	0.227769	0.089850	0.317619
False	True	True	0.010814	0.004528	0.015342
False	True	False	0.099848	0.041810	0.141658
False	False	True	0.025615	0.010104	0.035719
False	False	False	0.219997	0.086784	0.306781
Marginals			0.713400	0.286600	1.000000

Note: ZSMS = 0.000000 + 0.000000 = 0.000000
[a]From Tables 9.41, 9.42, 9.43, and 9.44, 0.010050 = 0.704853
* 0.481677 * 0.289345 * 0.102300

Step 9: Compute Posterior Probabilities

We compute the *posteriors* for $P(C2_j|D6_i, B1_i, A_i)$ and report this in Table 9.46.

Table 9.46 Posterior
probabilities

$P(C2_j|D6_i, B1_i, A_i)$					
$C2_j$					
$D6_i$	$B1_i$	A_i	True	False	Total
True	True	True	0.481677[a]	0.518323	1.000000
True	True	False	0.481677	0.518323	1.000000
True	False	True	0.508679	0.491321	1.000000
True	False	False	0.508679	0.491321	1.000000
False	True	True	0.481677	0.518323	1.000000
False	True	False	0.481677	0.518323	1.000000
False	False	True	0.508679	0.491321	1.000000
False	False	False	0.508679	0.491321	1.000000

[a]From Table 9.45, 0.481677 = 0.01005/(0.010050 + 0.010814)

We compute the *posteriors* for $P(B1_j|D6_i, C2_i, A_i)$ and report this in Table 9.47.

Table 9.47 Posterior probabilities

$P(B1_j \| D6_i, C2_i, A_i)$					
$B1_j$					
$D6_i$	$C2_i$	A_i	True	False	Total
True	True	True	0.274810[a]	0.725190	1.000000
True	True	False	0.289460	0.710540	1.000000
True	False	True	0.296857	0.703143	1.000000
True	False	False	0.312176	0.687824	1.000000
False	True	True	0.286860	0.713140	1.000000
False	True	False	0.301885	0.698115	1.000000
False	False	True	0.309461	0.690539	1.000000
False	False	False	0.325131	0.674869	1.000000

[a]From Table 9.45, $0.274810 = 0.01005/(0.010050 + 0.026520)$

We compute the *posteriors* for $P(A_j | D6_i, C2_i, B1_i)$ and report this in Table 9.48.

Table 9.48 Posterior probabilities

$P(A_j \| D6_i, C2_i, B1_i)$					
A_j					
$D6_i$	$C2_i$	$B1_i$	True	False	Total
True	True	True	0.097722[a]	0.902278	1.000000
True	True	False	0.104289	0.895711	1.000000
True	False	True	0.097722	0.902278	1.000000
True	False	False	0.104289	0.895711	1.000000
False	True	True	0.097722	0.902278	1.000000
False	True	False	0.104289	0.895711	1.000000
False	False	True	0.097722	0.902278	1.000000
False	False	False	0.104289	0.895711	1.000000

[a]From Table 9.45, $0.097722 = 0.01005/(0.010050 + 0.092789)$

9.4.7 Path 9: BBN Solution Protocol

Step 2: Specify the BBN–[$A \rightarrow B1 | B1 \rightarrow C2 | C2 \rightarrow D5$]

See Fig. 9.7.

Step 6: Determine Prior or Unconditional Probabilities

We select Node A for this BBN structure as reported in Table 9.49.

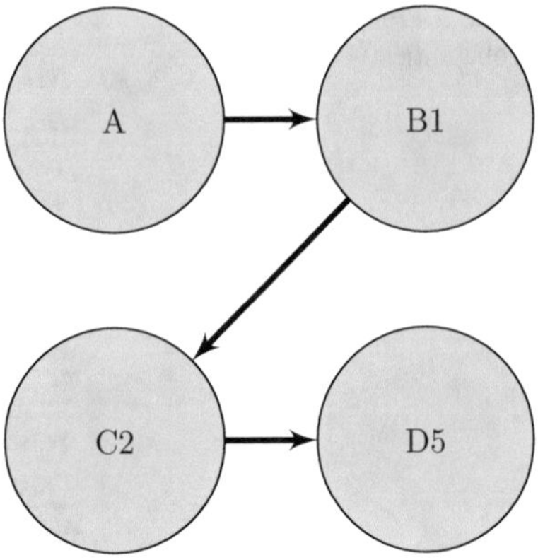

Fig. 9.7 Paths $[A \rightarrow B1|B1 \rightarrow C2|C2 \rightarrow D5]$

Table 9.49 Node A: prior probabilities

$P(A_j)$		
True	False	Total
0.102300[a]	0.897700	1.000000

[a]From Table 6.5, $0.102300 = 1023/10000$

Step 7: Determine Likelihood Probabilities

We select Node $B1$ for this BBN structure as reported in Table 9.50.

Table 9.50 Node $B1$: likelihood probabilities

$P(B1_j|A_i)$			
$B1_j$			
A_i	True	False	Total
True	0.289345[a]	0.710655	1.000000
False	0.304445	0.695555	1.000000

[a]From Table 6.8, $0.289345 = 296/1023$

We select Node $C2$ for this BBN structure as reported in Table 9.51.

Table 9.51 Node $C2$: likelihood probabilities

$P(C2_j|B1_i, A_i)$				
$C2_j$				
$B1_i$	A_i	True	False	Total
True	True	0.481677[a]	0.518323	1.000000
True	False	0.481677	0.518323	1.000000
False	True	0.508679	0.491321	1.000000
False	False	0.508679	0.491321	1.000000

[a]From Table 6.15, $0.481677 = 1459/3029$

We select Node $D5$ for this BBN structure as reported in Table 9.52.

Table 9.52 Node $D5$:
likelihood probabilities

$P(D5_j|C2_i, B1_i, A_i)$

			$D5_j$		
$C2_i$	$B1_i$	A_i	True	False	Total
True	True	True	0.710490[a]	0.289510	1.000000
True	True	False	0.710490	0.289510	1.000000
True	False	True	0.710490	0.289510	1.000000
True	False	False	0.710490	0.289510	1.000000
False	True	True	0.716316	0.283684	1.000000
False	True	False	0.716316	0.283684	1.000000
False	False	True	0.716316	0.283684	1.000000
False	False	False	0.716316	0.283684	1.000000

[a]From Table 6.30, $0.710490 = 3556/5005$

Step 8: Determine Joint and Marginal Probabilities

We compute the *joints* and *marginals* for $P(D5_j, C2_i, B1_i, A_i)$ and report these in Table 9.53.

Table 9.53 Joint and
marginal probabilities

$P(D5_j, C2_i, B1_i, A_i)$

			$D5_j$		
$C2_i$	$B1_i$	A_i	True	False	Marginals
True	True	True	0.010130[a]	0.004208	0.014258
True	True	False	0.093531	0.038112	0.131642
True	False	True	0.026275	0.010706	0.036981
True	False	False	0.225665	0.091954	0.317619
False	True	True	0.010990	0.004352	0.015342
False	True	False	0.101472	0.040186	0.141658
False	False	True	0.025586	0.010133	0.035719
False	False	False	0.219752	0.087029	0.306781
Marginals			0.713400	0.286600	1.000000

Note: ZSMS $= 0.000000 + 0.000000 = 0.000000$
[a]From Tables 9.49, 9.50, 9.51, and 9.52, $0.010130 = 0.710490$
$* 0.481677 * 0.289345 * 0.102300$

Step 9: Compute Posterior Probabilities

We compute the *posteriors* for $P(C2_j|D5_i, B1_i, A_i)$ and report this in Table 9.54.

Table 9.54 Posterior probabilities

$P(C2_j|D5_i, B1_i, A_i)$

$C2_j$					
$D5_i$	$B1_i$	A_i	True	False	Total
True	True	True	0.479638[a]	0.520362	1.000000
True	True	False	0.479638	0.520362	1.000000
True	False	True	0.506637	0.493363	1.000000
True	False	False	0.506637	0.493363	1.000000
False	True	True	0.486755	0.513245	1.000000
False	True	False	0.486755	0.513245	1.000000
False	False	True	0.513759	0.486241	1.000000
False	False	False	0.513759	0.486241	1.000000

[a]From Table 9.53, $0.479638 = 0.010130/(0.010130 + 0.010990)$

We compute the *posteriors* for $P(B1_j|D5_i, C2_i, A_i)$ and report this in Table 9.55.

Table 9.55 Posterior probabilities

$P(B1_j|D5_i, C2_i, A_i)$

$B1_j$					
$D5_i$	$C2_i$	A_i	True	False	Total
True	True	True	0.278260[a]	0.721740	1.000000
True	True	False	0.293020	0.706980	1.000000
True	False	True	0.300469	0.699531	1.000000
True	False	False	0.315891	0.684109	1.000000
False	True	True	0.278260	0.721740	1.000000
False	True	False	0.293020	0.706980	1.000000
False	False	True	0.300469	0.699531	1.000000
False	False	False	0.315891	0.684109	1.000000

[a]From Table 9.53, $0.278260 = 0.010130/(0.010130 + 0.026275)$

We compute the *posteriors* for $P(A_j|D5_i, C2_i, B1_i)$ and report this in Table 9.56.

9.4.8   Path 10: BBN Solution Protocol

Step 2: Specify the BBN–[$A \rightarrow B1|B1 \rightarrow D2|C4 \rightarrow D2$]

See Fig. 9.8.

Fig. 9.8 Paths [$A \rightarrow$ $B1|B1 \rightarrow D2|C4 \rightarrow D2$]

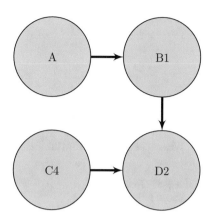

Table 9.56 Posterior probabilities

$P(A_j|D5_i, C2_i, B1_i)$

			A_j		
$D5_i$	$C2_i$	$B1_i$	True	False	Total
True	True	True	0.097722[a]	0.902278	1.000000
True	True	False	0.104289	0.895711	1.000000
True	False	True	0.097722	0.902278	1.000000
True	False	False	0.104289	0.895711	1.000000
False	True	True	0.097722	0.902278	1.000000
False	True	False	0.104289	0.895711	1.000000
False	False	True	0.097722	0.902278	1.000000
False	False	False	0.104289	0.895711	1.000000

[a]From Table 9.53, $0.097722 = 0.010130/(0.010130 + 0.093531)$

Step 6: Determine Prior or Unconditional Probabilities

We select Node A for this BBN structure as reported in Table 9.57.

Table 9.57 Node A: prior probabilities

$P(A_j)$

True	False	Total
0.102300[a]	0.897700	1.000000

[a]From Table 6.5, $0.102300 = 1023/10000$

Step 7: Determine Likelihood Probabilities

We select Node $B1$ for this BBN structure as reported in Table 9.58.

Table 9.58 Node $B1$:
likelihood probabilities

$P(B1_j|A_i)$			
$B1_j$			
A_i	True	False	Total
True	0.289345[a]	0.710655	1.000000
False	0.304445	0.695555	1.000000

[a]From Table 6.8, 0.289345 = 296/1023

We select Node $C4$ for this BBN structure as reported in Table 9.59.

Table 9.59 Node $C4$:
likelihood probabilities

$P(C4_j|B1_i, A_i)$					
$C4_j$					
$B1_i$	A_i	True	False	Total
~~True~~	~~True~~	0.500500[a]	0.499500	1.000000
~~True~~	~~False~~	~~0.500500~~	~~0.499500~~	~~1.000000~~
~~False~~	~~True~~	~~0.500500~~	~~0.499500~~	~~1.000000~~
~~False~~	~~False~~	~~0.500500~~	~~0.499500~~	~~1.000000~~

[a]From Table 6.19, 0.500500 = 5005/10000

We select Node $D2$ for this BBN structure as reported in Table 9.60.

Table 9.60 Node $D2$:
likelihood probabilities

$P(D2_j|C4_i, B1_i, A_i)$					
$D2_j$					
$C4_i$	$B1_i$	A_i	True	False	Total
True	True	~~True~~	0.692940[a]	0.307060	1.000000
~~True~~	~~True~~	~~False~~	~~0.692940~~	~~0.307060~~	~~1.000000~~
True	False	~~True~~	0.717710	0.282290	1.000000
~~True~~	~~False~~	~~False~~	~~0.717710~~	~~0.282290~~	~~1.000000~~
False	True	~~True~~	0.715924	0.284076	1.000000
~~False~~	~~True~~	~~False~~	~~0.715924~~	~~0.284076~~	~~1.000000~~
False	False	~~True~~	0.716496	0.283504	1.000000
~~False~~	~~False~~	~~False~~	~~0.716496~~	~~0.283504~~	~~1.000000~~

[a]From Table 6.24, 0.692940 = 1011/1459

Step 8: Determine Joint and Marginal Probabilities

We compute the *joints* and *marginals* for $P(D2_j, C4_i, B1_i, A_i)$ and report these in Table 9.61.

Table 9.61 Joint and marginal probabilities

$P(D2_j, C4_i, B1_i, A_i)$					
$D2_j$					
$C4_i$	$B1_i$	A_i	True	False	Marginals
True	True	True	0.010266[a]	0.004549	0.014815
True	True	False	0.094785	0.042002	0.136787
True	False	True	0.026115	0.010271	0.036386
True	False	False	0.224293	0.088219	0.312512
False	True	True	0.010585	0.004200	0.014785
False	True	False	0.097733	0.038780	0.136513
False	False	True	0.026019	0.010295	0.036314
False	False	False	0.223466	0.088421	0.311888
Marginals			0.713262	0.286738	1.000000

Note: ZSMS = 0.000138 − 0.000138 = 0.000000
[a]From Tables 9.57, 9.58, 9.59, and 9.60, 0.010266 = 0.692940 * 0.500500 * 0.289345 * 0.102300

Step 9: Compute Posterior Probabilities

We compute the *posteriors* for $P(C4_j|D2_i, B1_i, A_i)$ and report this in Table 9.62.

Table 9.62 Posterior probabilities

$P(C4_j|D2_i, B1_i, A_i)$					
$C4_j$					
$D2_i$	$B1_i$	A_i	True	False	Total
True	True	True	0.492343[a]	0.507657	1.000000
True	True	False	0.492343	0.507657	1.000000
True	False	True	0.500923	0.499077	1.000000
True	False	False	0.500923	0.499077	1.000000
False	True	True	0.519939	0.480061	1.000000
False	True	False	0.519939	0.480061	1.000000
False	False	True	0.499427	0.500573	1.000000
False	False	False	0.499427	0.500573	1.000000

[a]From Table 9.61, 0.492343 = 0.010266/(0.010266 + 0.010585)

We compute the *posteriors* for $P(B1_j|D2_i, C4_i, A_i)$ and report this in Table 9.63.

Table 9.63 Posterior
probabilities

$P(B1_j \mid D2_i, C4_i, A_i)$					
$B1_j$					
$D2_i$	$C4_i$	A_i	True	False	Total
True	True	True	0.282177^a	0.717823	1.000000
True	True	False	0.297059	0.702941	1.000000
True	False	True	0.289181	0.710819	1.000000
True	False	False	0.304275	0.695725	1.000000
False	True	True	0.306941	0.693059	1.000000
False	True	False	0.322542	0.677458	1.000000
False	False	True	0.289760	0.710240	1.000000
False	False	False	0.304872	0.695128	1.000000

[a] From Table 9.61, $0.282177 = 0.010266/(0.010266 + 0.026115)$

We compute the *posteriors* for $P(A_j \mid D2_i, C4_i, B1_i)$ and report this in Table 9.64.

Table 9.64 Posterior
probabilities

$P(A_j \mid D2_i, C4_i, B1_i)$					
A_j					
$D2_i$	$C4_i$	$B1_i$	True	False	Total
True	True	True	0.097722^a	0.902278	1.000000
True	True	False	0.104289	0.895711	1.000000
True	False	True	0.097722	0.902278	1.000000
True	False	False	0.104289	0.895711	1.000000
False	True	True	0.097722	0.902278	1.000000
False	True	False	0.104289	0.895711	1.000000
False	False	True	0.097722	0.902278	1.000000
False	False	False	0.104289	0.895711	1.000000

[a] From Table 9.61, $0.097722 = 0.010266/(0.010266 + 0.094785)$

9.4.9 Path 11: BBN Solution Protocol

Step 2: Specify the BBN–[A → C1|A → D7|B2 → C1]

See Fig. 9.9.

Step 6: Determine Prior or Unconditional Probabilities

We select Node A for this BBN structure as reported in Table 9.65.

Fig. 9.9 Paths $[A \rightarrow C1|A \rightarrow D7|B2 \rightarrow C1]$

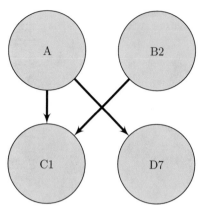

Table 9.65 Node A: prior probabilities

$P(A_j)$		
True	False	Total
0.102300[a]	0.897700	1.000000

[a]From Table 6.5, 0.102300 = 1023/10000

Step 7: Determine Likelihood Probabilities

We select Node $B2$ for this BBN structure as reported in Table 9.66.

Table 9.66 Node $B2$: likelihood probabilities

$P(B2_j|A_i)$			
$B2_j$			
A_i	True	False	Total
~~True~~	0.302900[a]	0.697100	1.000000
~~False~~	~~0.302900~~	~~0.697100~~	~~1.000000~~

[a]From Table 6.10, 0.302900 = 3029/10000

We select Node $C1$ for this BBN structure as reported in Table 9.67.

Table 9.67 Node $C1$: likelihood probabilities

$P(C1_j|B2_i, A_i)$				
$C1_j$				
$B2_i$	A_i	True	False	Total
True	True	0.466216[a]	0.533784	1.000000
True	False	0.483352	0.516648	1.000000
False	True	0.478680	0.521320	1.000000
False	False	0.512172	0.487828	1.000000

[a]From Table 6.13, 0.466216 = 138/296

We select Node $D7$ for this BBN structure as reported in Table 9.68.

Table 9.68 Node $D7$:
likelihood probabilities

$P(D7_j | C1_i, B2_i, A_i)$

$D7_j$					
$C1_i$	$B2_i$	A_i	True	False	Total
True	True	True	0.730205[a]	0.269795	1.000000
True	True	False	0.711485	0.288515	1.000000
True	False	True	0.730205	0.269795	1.000000
True	False	False	0.711485	0.288515	1.000000
False	True	True	0.730205	0.269795	1.000000
False	True	False	0.711485	0.288515	1.000000
False	False	True	0.730205	0.269795	1.000000
False	False	False	0.711485	0.288515	1.000000

[a]From Table 6.34, 0.730205 = 747/1023

Step 8: Determine Joint and Marginal Probabilities

We compute the *joints* and *marginals* for $P(D7_j, C1_i, B2_i, A_i)$ and report these in Table 9.69.

Table 9.69 Joint and
marginal probabilities

$P(D7_j, C1_i, B2_i, A_i)$

$D7_j$					
$C1_i$	$B2_i$	A_i	True	False	Marginals
True	True	True	0.010549[a]	0.003898	0.014446
True	True	False	0.093510	0.037919	0.131430
True	False	True	0.024926	0.009210	0.034136
True	False	False	0.228038	0.092472	0.320510
False	True	True	0.012078	0.004462	0.016540
False	True	False	0.099952	0.040532	0.140484
False	False	True	0.027147	0.010030	0.037177
False	False	False	0.217200	0.088077	0.305276
Marginals			0.713400	0.286600	1.000000

Note: ZSMS = 0.000000 + 0.000000 = 0.000000
[a]From Tables 9.65, 9.66, 9.67, and 9.68, 0.010549 = 0.730205
* 0.466216 * 0.302900 * 0.102300

Step 9: Compute Posterior Probabilities

We compute the *posteriors* for $P(C1_j | D7_i, B2_i, A_i)$ and report this in Table 9.70.

Table 9.70 Posterior probabilities

$P(C1_j\|D7_i, B2_i, A_i)$					
$C1_j$					
$D7_i$	$B2_i$	A_i	True	False	Total
True	True	True	0.466216[a]	0.533784	1.000000
True	True	False	0.483352	0.516648	1.000000
True	False	True	0.478680	0.521320	1.000000
True	False	False	0.512172	0.487828	1.000000
False	True	True	0.466216	0.533784	1.000000
False	True	False	0.483352	0.516648	1.000000
False	False	True	0.478680	0.521320	1.000000
False	False	False	0.512172	0.487828	1.000000

[a]From Table 9.69, $0.466216 = 0.010549/(0.010549 + 0.012078)$

We compute the *posteriors* for $P(B2_j|D7_i, C1_i, A_i)$ and report this in Table 9.71.

Table 9.71 Posterior probabilities

$P(B2_j\|D7_i, C1_i, A_i)$					
$B2_j$					
$D7_i$	$C1_i$	A_i	True	False	Total
True	True	True	0.297359[a]	0.702641	1.000000
True	True	False	0.290812	0.709188	1.000000
True	False	True	0.307912	0.692088	1.000000
True	False	False	0.315155	0.684845	1.000000
False	True	True	0.297359	0.702641	1.000000
False	True	False	0.290812	0.709188	1.000000
False	False	True	0.307912	0.692088	1.000000
False	False	False	0.315155	0.684845	1.000000

[a]From Table 9.69, $0.297359 = 0.010549/(0.010549 + 0.024926)$

We compute the *posteriors* for $P(A_j|D7_i, C1_i, B2_i)$ and report this in Table 9.72.

9.4.10 Path 12: BBN Solution Protocol

Step 2: Specify the BBN–$[A \rightarrow C3|A \rightarrow D4|B2 \rightarrow D4]$

See Fig. 9.10.

Table 9.72 Posterior probabilities

$P(A_j|D7_i, C1_i, B2_i)$					
A_j					
$D7_i$	$C1_i$	$B2_i$	True	False	Total
True	True	True	0.101374[a]	0.898626	1.000000
True	True	False	0.098537	0.901463	1.000000
True	False	True	0.107808	0.892192	1.000000
True	False	False	0.111100	0.888900	1.000000
False	True	True	0.093206	0.906794	1.000000
False	True	False	0.090574	0.909426	1.000000
False	False	True	0.099179	0.900821	1.000000
False	False	False	0.102237	0.897763	1.000000

[a]From Table 9.69, $0.101374 = 0.010549/(0.010549 + 0.093510)$

Fig. 9.10 Paths $[A \rightarrow C3|A \rightarrow D4|B2 \rightarrow D4]$

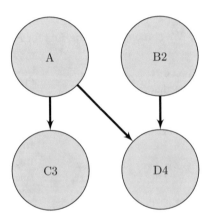

Step 6: Determine Prior or Unconditional Probabilities

We select Node A for this BBN structure as reported in Table 9.73.

Table 9.73 Node A: prior probabilities

$P(A_j)$		
True	False	Total
0.102300[a]	0.897700	1.000000

[a]From Table 6.5, $0.102300 = 1023/10000$

Step 7: Determine Likelihood Probabilities

We select Node $B2$ for this BBN structure as reported in Table 9.74.
We select Node $C3$ for this BBN structure as reported in Table 9.75.

Table 9.74 Node $B2$: likelihood probabilities

$P(B2_j\|A_i)$			
$B2_j$			
A_i	True	False	Total
True	0.302900[a]	0.697100	1.000000
False	0.302900	0.697100	1.000000

[a]From Table 6.10, $0.302900 = 3029/10000$

Table 9.75 Node $C3$: likelihood probabilities

$P(C3_j\|B2_i, A_i)$				
$C3_j$				
$B2_i$	A_i	True	False	Total
True	True	0.475073[a]	0.524927	1.000000
True	False	0.503398	0.496602	1.000000
False	True	0.475073	0.524927	1.000000
False	False	0.503398	0.496602	1.000000

[a]From Table 6.17, $0.475073 = 486/1023$

We select Node $D4$ for this BBN structure as reported in Table 9.76.

Table 9.76 Node $D4$: likelihood probabilities

$P(D4_j\|C3_i, B2_i, A_i)$					
$D4_j$					
$C3_i$	$B2_i$	A_i	True	False	Total
True	True	True	0.685811[a]	0.314189	1.000000
True	True	False	0.706915	0.293085	1.000000
True	False	True	0.748281	0.251719	1.000000
True	False	False	0.713485	0.286515	1.000000
False	True	True	0.685811	0.314189	1.000000
False	True	False	0.706915	0.293085	1.000000
False	False	True	0.748281	0.251719	1.000000
False	False	False	0.713485	0.286515	1.000000

[a]From Table 6.28, $0.685811 = 203/296$

Step 8: Determine Joint and Marginal Probabilities

We compute the *joints* and *marginals* for $P(D4_j, C3_i, B2_i, A_i)$ and report these in Table 9.77.

Step 9: Compute Posterior Probabilities

We compute the *posteriors* for $P(C3_j|D4_i, B2_i, A_i)$ and report this in Table 9.78.

Table 9.77 Joint and marginal probabilities

$P(D4_j, C3_i, B2_i, A_i)$

$C3_i$	$B2_i$	A_i	$D4_j$ True	False	Marginals
True	True	True	0.010096[a]	0.004625	0.014721
True	True	False	0.096763	0.040118	0.136881
True	False	True	0.025351	0.008528	0.033879
True	False	False	0.224762	0.090258	0.315019
False	True	True	0.011155	0.005111	0.016266
False	True	False	0.095457	0.039576	0.135033
False	False	True	0.028011	0.009423	0.037434
False	False	False	0.221728	0.089039	0.310767
Marginals			0.713322	0.286678	1.000000

Note: ZSMS = 0.000078 − 0.000078 = 0.000000
[a]From Tables 9.73, 9.74, 9.75, and 9.76, 0.010096 = 0.685811 * 0.475073 * 0.302900 * 0.102300

Table 9.78 Posterior probabilities

$P(C3_j | D4_i, B2_i, A_i)$

$D4_i$	$B2_i$	A_i	$C3_j$ True	False	Total
True	True	True	0.475073[a]	0.524927	1.000000
True	True	False	0.503398	0.496602	1.000000
True	False	True	0.475073	0.524927	1.000000
True	False	False	0.503398	0.496602	1.000000
False	True	True	0.475073	0.524927	1.000000
False	True	False	0.503398	0.496602	1.000000
False	False	True	0.475073	0.524927	1.000000
False	False	False	0.503398	0.496602	1.000000

[a]From Table 9.77, 0.475073 = 0.010096/(0.010096 + 0.011155)

We compute the *posteriors* for $P(B2_j | D4_i, C3_i, A_i)$ and report this in Table 9.79. We compute the *posteriors* for $P(A_j | D4_i, C3_i, B2_i)$ and report this in Table 9.80.

9.4.11 Path 14: BBN Solution Protocol

Step 2: Specify the BBN–[A → C1|B2 → C1|B2 → D6]

See Fig. 9.11.

Table 9.79 Posterior
probabilities

$P(B2_j|D4_i, C3_i, A_i)$					
$B2_j$					
$D4_i$	$C3_i$	A_i	True	False	Total
True	True	True	0.284815[a]	0.715185	1.000000
True	True	False	0.300950	0.699050	1.000000
True	False	True	0.284815	0.715185	1.000000
True	False	False	0.300950	0.699050	1.000000
False	True	True	0.351638	0.648362	1.000000
False	True	False	0.307708	0.692292	1.000000
False	False	True	0.351638	0.648362	1.000000
False	False	False	0.307708	0.692292	1.000000

[a]From Table 9.77, 0.284815 $=$ 0.010096/(0.010096 $+$ 0.025351)

Table 9.80 Posterior
probabilities

$P(A_j|D4_i, C3_i, B2_i)$					
A_j					
$D4_i$	$C3_i$	$B2_i$	True	False	Total
True	True	True	0.094478[a]	0.905522	1.000000
True	True	False	0.101358	0.898642	1.000000
True	False	True	0.104634	0.895366	1.000000
True	False	False	0.112162	0.887838	1.000000
False	True	True	0.103372	0.896628	1.000000
False	True	False	0.086328	0.913672	1.000000
False	False	True	0.114364	0.885636	1.000000
False	False	False	0.095701	0.904299	1.000000

[a]From Table 9.77, 0.094478 $=$ 0.010096/(0.010096 $+$ 0.096763)

Fig. 9.11 Paths $[A \rightarrow C1|B2 \rightarrow C1|B2 \rightarrow D6]$

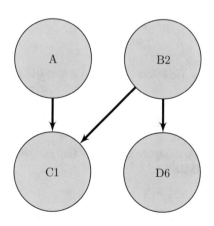

Step 6: Determine Prior or Unconditional Probabilities

We select Node A for this BBN structure as reported in Table 9.81.

Table 9.81 Node A: prior
probabilities

$P(A_j)$		
True	False	Total
0.102300[a]	0.897700	1.000000

[a]From Table 6.5, 0.102300 =
1023/10000

Step 7: Determine Likelihood Probabilities

We select Node $B2$ for this BBN structure as reported in Table 9.82.

Table 9.82 Node $B2$:
likelihood probabilities

$P(B2_j|A_i)$			
$B2_j$			
A_i	True	False	Total
~~True~~	0.302900[a]	0.697100	1.000000
~~False~~	~~0.302900~~	~~0.697100~~	~~1.000000~~

[a]From Table 6.10, 0.302900 = 3029/10000

We select Node $C1$ for this BBN structure as reported in Table 9.83.

Table 9.83 Node $C1$:
likelihood probabilities

$P(C1_j|B2_i, A_i)$				
$C1_j$				
$B2_i$	A_i	True	False	Total
True	True	0.466216[a]	0.533784	1.000000
True	False	0.483352	0.516648	1.000000
False	True	0.478680	0.521320	1.000000
False	False	0.512172	0.487828	1.000000

[a]From Table 6.13, 0.466216 = 138/296

We select Node $D6$ for this BBN structure as reported in Table 9.84.

Step 8: Determine Joint and Marginal Probabilities

We compute the *joints* and *marginals* for $P(D6_j, C1_i, B2_i, A_i)$ and report these in
Table 9.85.

Step 9: Compute Posterior Probabilities

We compute the *posteriors* for $P(C1_j|D6_i, B2_i, A_i)$ and report this in Table 9.86.
We compute the *posteriors* for $P(B2_j|D6_i, C1_i, A_i)$ and report this in Table 9.87.
We compute the *posteriors* for $P(A_j|D6_i, C1_i, B2_i)$ and report this in Table 9.88.

Table 9.84 Node $D6$:
likelihood probabilities

$P(D6_j | C1_i, B2_i, A_i)$

$C1_i$	$B2_i$	A_i	$D6_j$ True	False	Total
True	True	True	0.704853[a]	0.295147	1.000000
True	True	False	0.704853	0.295147	1.000000
True	False	True	0.717114	0.282886	1.000000
True	False	False	0.717114	0.282886	1.000000
False	True	True	0.704853	0.295147	1.000000
False	True	False	0.704853	0.295147	1.000000
False	False	True	0.717114	0.282886	1.000000
False	False	False	0.717114	0.282886	1.000000

[a]From Table 6.32, 0.704853 = 2135/3029

Table 9.85 Joint and
marginal probabilities

$P(D6_j, C1_i, B2_i, A_i)$

$C1_i$	$B2_i$	A_i	$D6_j$ True	False	Marginals
True	True	True	0.010183[a]	0.004264	0.014446
True	True	False	0.092639	0.038791	0.131430
True	False	True	0.024480	0.009657	0.034136
True	False	False	0.229842	0.090668	0.320510
False	True	True	0.011658	0.004882	0.016540
False	True	False	0.099020	0.041463	0.140484
False	False	True	0.026660	0.010517	0.037177
False	False	False	0.218918	0.086359	0.305276
Marginals			0.713400	0.286600	1.000000

Note: ZSMS = 0.000000 + 0.000000 = 0.000000
[a]From Tables 9.81, 9.82, 9.83, and 9.84, 0.010183 = 0.704853 * 0.466216 * 0.302900 * 0.102300

Table 9.86 Posterior
probabilities

$P(C1_j | D6_i, B2_i, A_i)$

$D6_i$	$B2_i$	A_i	$C1_j$ True	False	Total
True	True	True	0.466216[a]	0.533784	1.000000
True	True	False	0.483352	0.516648	1.000000
True	False	True	0.478680	0.521320	1.000000
True	False	False	0.512172	0.487828	1.000000
False	True	True	0.466216	0.533784	1.000000
False	True	False	0.483352	0.516648	1.000000
False	False	True	0.478680	0.521320	1.000000
False	False	False	0.512172	0.487828	1.000000

[a]From Table 9.85, 0.466216 = 0.010183/(0.010183 + 0.011658)

Table 9.87 Posterior
probabilities

$P(B2_j|D6_i, C1_i, A_i)$					
$B2_j$					
$D6_i$	$C1_i$	A_i	True	False	Total
True	True	True	0.293768[a]	0.706232	1.000000
True	True	False	0.287269	0.712731	1.000000
True	False	True	0.304249	0.695751	1.000000
True	False	False	0.311445	0.688555	1.000000
False	True	True	0.306299	0.693701	1.000000
False	True	False	0.299640	0.700360	1.000000
False	False	True	0.317026	0.682974	1.000000
False	False	False	0.324384	0.675616	1.000000

[a]From Table 9.85, $0.293768 = 0.010183/(0.010183 + 0.024480)$

Table 9.88 Posterior
probabilities

$P(A_j|D6_i, C1_i, B2_i)$					
A_j					
$D6_i$	$C1_i$	$B2_i$	True	False	Total
True	True	True	0.099032[a]	0.900968	1.000000
True	True	False	0.096254	0.903746	1.000000
True	False	True	0.105336	0.894664	1.000000
True	False	False	0.108561	0.891439	1.000000
False	True	True	0.099032	0.900968	1.000000
False	True	False	0.096254	0.903746	1.000000
False	False	True	0.105336	0.894664	1.000000
False	False	False	0.108561	0.891439	1.000000

[a]From Table 9.85, $0.099032 = 0.010183/(0.010183 + 0.092639)$

9.4.12 Path 15: BBN Solution Protocol

Step 2: Specify the BBN–[$A \rightarrow C1|B2 \rightarrow C1|C1 \rightarrow D5$]

See Fig. 9.12.

Step 6: Determine Prior or Unconditional Probabilities

We select Node A for this BBN structure as reported in Table 9.89.

Step 7: Determine Likelihood Probabilities

We select Node $B2$ for this BBN structure as reported in Table 9.90.
We select Node $C1$ for this BBN structure as reported in Table 9.91.

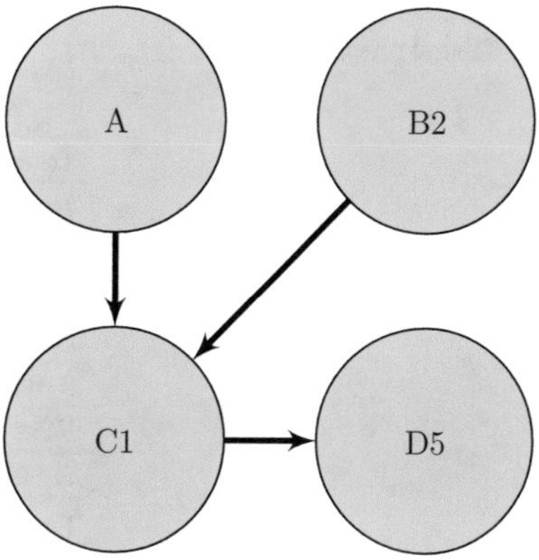

Fig. 9.12 Paths $[A \rightarrow C1|B2 \rightarrow C1|C1 \rightarrow D5]$

Table 9.89 Node A: prior probabilities

$P(A_j)$		
True	False	Total
0.102300[a]	0.897700	1.000000

[a]From Table 6.5, $0.102300 = 1023/10000$

Table 9.90 Node $B2$: likelihood probabilities

$P(B2_j|A_i)$			
$B2_j$			
A_i	True	False	Total
True	0.302900[a]	0.697100	1.000000
False	0.302900	0.697100	1.000000

[a]From Table 6.10, $0.302900 = 3029/10000$

Table 9.91 Node $C1$: likelihood probabilities

$P(C1_j|B2_i, A_i)$				
$C1_j$				
$B2_i$	A_i	True	False	Total
True	True	0.466216[a]	0.533784	1.000000
True	False	0.483352	0.516648	1.000000
False	True	0.478680	0.521320	1.000000
False	False	0.512172	0.487828	1.000000

[a]From Table 6.13, $0.466216 = 138/296$

We select Node $D5$ for this BBN structure as reported in Table 9.92.

Step 8: Determine Joint and Marginal Probabilities

We compute the *joints* and *marginals* for $P(D5_j, C1_i, B2_i, A_i)$ and report these in Table 9.93.

Table 9.92 Node $D5$: likelihood probabilities

$P(D5_j | C1_i, \cancel{B2_i}, \cancel{A_i})$

$D5_j$

$C1_i$	$\cancel{B2_i}$	$\cancel{A_i}$	True	False	Total
True	~~True~~	~~True~~	0.710490[a]	0.289510	1.000000
~~True~~	~~True~~	~~False~~	~~0.710490~~	~~0.289510~~	~~1.000000~~
~~True~~	~~False~~	~~True~~	~~0.710490~~	~~0.289510~~	~~1.000000~~
~~True~~	~~False~~	~~False~~	~~0.710490~~	~~0.289510~~	~~1.000000~~
False	~~True~~	~~True~~	0.716316	0.283684	1.000000
~~False~~	~~True~~	~~False~~	~~0.716316~~	~~0.283684~~	~~1.000000~~
~~False~~	~~False~~	~~True~~	~~0.716316~~	~~0.283684~~	~~1.000000~~
~~False~~	~~False~~	~~False~~	~~0.716316~~	~~0.283684~~	~~1.000000~~

[a]From Table 6.30, 0.710490 = 3556/5005

Table 9.93 Joint and marginal probabilities

$P(D5_j, C1_i, B2_i, A_i)$

$D5_j$

$C1_i$	$B2_i$	A_i	True	False	Marginals
True	True	True	0.010264[a]	0.004182	0.014446
True	True	False	0.093379	0.038050	0.131430
True	False	True	0.024253	0.009883	0.034136
True	False	False	0.227719	0.092791	0.320510
False	True	True	0.011848	0.004692	0.016540
False	True	False	0.100631	0.039853	0.140484
False	False	True	0.026631	0.010547	0.037177
False	False	False	0.218675	0.086602	0.305276
Marginals			0.713400	0.286600	1.000000

Note: ZSMS $= 0.000000 + 0.000000 = 0.000000$
[a]From Tables 9.89, 9.90, 9.91, and 9.92, 0.010264 = 0.710490 * 0.466216 * 0.302900 * 0.102300

Step 9: Compute Posterior Probabilities

We compute the *posteriors* for $P(C1_j | D5_i, B2_i, A_i)$ and report this in Table 9.94.
We compute the *posteriors* for $P(B2_j | D5_i, C1_i, A_i)$ and report this in Table 9.95.
We compute the *posteriors* for $P(A_j | D5_i, C1_i, B2_i)$ and report this in Table 9.96.

9.4.13   Path 16: BBN Solution Protocol

Step 2: Specify the BBN–[$A \rightarrow C3 | B2 \rightarrow D2 | C3 \rightarrow D2$]

See Fig. 9.13.

Table 9.94 Posterior
probabilities

$P(C1_j\|D5_i, B2_i, A_i)$					
$C1_j$					
$D5_i$	$B2_i$	A_i	True	False	Total
True	True	True	0.464184[a]	0.535816	1.000000
True	True	False	0.481312	0.518688	1.000000
True	False	True	0.476642	0.523358	1.000000
True	False	False	0.510131	0.489869	1.000000
False	True	True	0.471279	0.528721	1.000000
False	True	False	0.488430	0.511570	1.000000
False	False	True	0.483755	0.516245	1.000000
False	False	False	0.517250	0.482750	1.000000

[a]From Table 9.93, 0.464184 = 0.010264/(0.010264 + 0.011848)

Table 9.95 Posterior
probabilities

$P(B2_j\|D5_i, C1_i, A_i)$					
$B2_j$					
$D5_i$	$C1_i$	A_i	True	False	Total
True	True	True	0.297359[a]	0.702641	1.000000
True	True	False	0.290812	0.709188	1.000000
True	False	True	0.307912	0.692088	1.000000
True	False	False	0.315155	0.684845	1.000000
False	True	True	0.297359	0.702641	1.000000
False	True	False	0.290812	0.709188	1.000000
False	False	True	0.307912	0.692088	1.000000
False	False	False	0.315155	0.684845	1.000000

[a]From Table 9.93, 0.297359 = 0.010264/(0.010264 + 0.024253)

Table 9.96 Posterior
probabilities

$P(A_j\|D5_i, C1_i, B2_i)$					
A_j					
$D5_i$	$C1_i$	$B2_i$	True	False	Total
True	True	True	0.099032[a]	0.900968	1.000000
True	True	False	0.096254	0.903746	1.000000
True	False	True	0.105336	0.894664	1.000000
True	False	False	0.108561	0.891439	1.000000
False	True	True	0.099032	0.900968	1.000000
False	True	False	0.096254	0.903746	1.000000
False	False	True	0.105336	0.894664	1.000000
False	False	False	0.108561	0.891439	1.000000

[a]From Table 9.93, 0.099032 = 0.010264/(0.010264 + 0.093379)

Step 6: Determine Prior or Unconditional Probabilities

We select Node *A* for this BBN structure as reported in Table 9.97.

Fig. 9.13 Paths [$A \to$ $C3|B2 \to D2|C3 \to D2$]

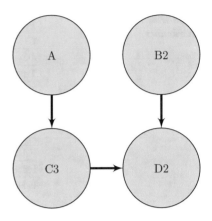

Table 9.97 Node A: prior probabilities

$P(A_j)$		
True	False	Total
0.102300[a]	0.897700	1.000000

[a]From Table 6.5, 0.102300 = 1023/10000

Step 7: Determine Likelihood Probabilities

We select Node $B2$ for this BBN structure as reported in Table 9.98

Table 9.98 Node $B2$: likelihood probabilities

$P(B2_j|A_i)$			
$B2_j$			
A_i	True	False	Total
True	0.302900[a]	0.697100	1.000000
False	0.302900	0.697100	1.000000

[a]From Table 6.10, 0.302900 = 3029/10000

We select Node $C3$ for this BBN structure as reported in Table 9.99.

Table 9.99 Node $C3$: likelihood probabilities

$P(C3_j|B2_i, A_i)$				
$C3_j$				
$B2_i$	A_i	True	False	Total
True	True	0.475073[a]	0.524927	1.000000
True	False	0.503398	0.496602	1.000000
False	True	0.475073	0.524927	1.000000
False	False	0.503398	0.496602	1.000000

[a]From Table 6.17, 0.475073 = 486/1023

We select Node $D2$ for this BBN structure as reported in Table 9.100.

Table 9.100 Node $D2$:
likelihood probabilities

$P(D2_j\|C3_i, B2_i, A_i)$					
$D2_j$					
$C3_i$	$B2_i$	A_i	True	False	Total
True	True	True	0.692940[a]	0.307060	1.000000
True	True	False	0.692940	0.307060	1.000000
True	False	True	0.717710	0.282290	1.000000
True	False	False	0.717710	0.282290	1.000000
False	True	True	0.715924	0.284076	1.000000
False	True	False	0.715924	0.284076	1.000000
False	False	True	0.716496	0.283504	1.000000
False	False	False	0.716496	0.283504	1.000000

[a]From Table 6.24, $0.692940 = 1011/1459$

Step 8: Determine Joint and Marginal Probabilities

We compute the *joints* and *marginals* for $P(D2_j, C3_i, B2_i, A_i)$ and report these in
Table 9.101.

Table 9.101 Joint and
marginal probabilities

$P(D2_j, C3_i, B2_i, A_i)$					
$D2_j$					
$C3_i$	$B2_i$	A_i	True	False	Marginals
True	True	True	0.010201[a]	0.004520	0.014721
True	True	False	0.094850	0.042030	0.136881
True	False	True	0.024315	0.009564	0.033879
True	False	False	0.226093	0.088927	0.315019
False	True	True	0.011645	0.004621	0.016266
False	True	False	0.096673	0.038360	0.135033
False	False	True	0.026822	0.010613	0.037434
False	False	False	0.222664	0.088104	0.310767
Marginals			0.713262	0.286738	1.000000

Note: ZSMS $= 0.000138 - 0.000138 = 0.000000$
[a]From Tables 9.97, 9.98, 9.99, and 9.100, $0.010201 =$
$0.692940 * 0.475073 * 0.302900 * 0.102300$

Step 9: Compute Posterior Probabilities

We compute the *posteriors* for $P(C3_j|D2_i, B2_i, A_i)$ and report this in Table 9.102.
We compute the *posteriors* for $P(B2_j|D2_i, C3_i, A_i)$ and report this in Table 9.103.
We compute the *posteriors* for $P(A_j|D2_i, C3_i, B2_i)$ and report this in Table 9.104.

Table 9.102 Posterior probabilities

$P(C3_j|D2_i, B2_i, A_i)$

$C3_j$					
$D2_i$	$B2_i$	A_i	True	False	Total
True	True	True	0.466944[a]	0.533056	1.000000
True	True	False	0.495240	0.504760	1.000000
True	False	True	0.475495	0.524505	1.000000
True	False	False	0.503821	0.496179	1.000000
False	True	True	0.494503	0.505497	1.000000
False	True	False	0.522831	0.477169	1.000000
False	False	True	0.474003	0.525997	1.000000
False	False	False	0.502325	0.497675	1.000000

[a]From Table 9.101, 0.466944 = 0.010201/(0.010201 + 0.011645)

Table 9.103 Posterior probabilities

$P(B2_j|D2_i, C3_i, A_i)$

$B2_j$					
$D2_i$	$C3_i$	A_i	True	False	Total
True	True	True	0.295536[a]	0.704464	1.000000
True	True	False	0.295536	0.704464	1.000000
True	False	True	0.302731	0.697269	1.000000
True	False	False	0.302731	0.697269	1.000000
False	True	True	0.320948	0.679052	1.000000
False	True	False	0.320948	0.679052	1.000000
False	False	True	0.303326	0.696674	1.000000
False	False	False	0.303326	0.696674	1.000000

[a]From Table 9.101, 0.295536 = 0.010201/(0.010201 + 0.024315)

Table 9.104 Posterior probabilities

$P(A_j|D2_i, C3_i, B2_i)$

A_j					
$D2_i$	$C3_i$	$B2_i$	True	False	Total
True	True	True	0.097103[a]	0.902897	1.000000
True	True	False	0.097103	0.902897	1.000000
True	False	True	0.107508	0.892492	1.000000
True	False	False	0.107508	0.892492	1.000000
False	True	True	0.097103	0.902897	1.000000
False	True	False	0.097103	0.902897	1.000000
False	False	True	0.107508	0.892492	1.000000
False	False	False	0.107508	0.892492	1.000000

[a]From Table 9.101, 0.097103 = 0.010201/(0.010201 + 0.094850)

Fig. 9.14 Paths [$A \rightarrow$
$D4|B2 \rightarrow C2|B2 \rightarrow D4$]

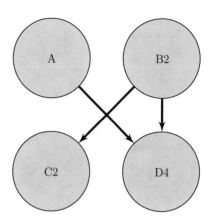

9.4.14   Path 17: BBN Solution Protocol

Step 2: Specify the BBN–[$A \rightarrow D4|B2 \rightarrow C2|B2 \rightarrow D4$]

See Fig. 9.14.

Step 6: Determine Prior or Unconditional Probabilities

We select Node A for this BBN structure as reported in Table 9.105.

Table 9.105 Node A: prior
probabilities

$P(A_j)$		
True	False	Total
0.102300[a]	0.897700	1.000000

[a]From Table 6.5, 0.102300 $=$ 1023/10000

Step 7: Determine Likelihood Probabilities

We select Node $B2$ for this BBN structure as reported in Table 9.106.
We select Node $C2$ for this BBN structure as reported in Table 9.107.
We select Node $D4$ for this BBN structure as reported in Table 9.108.

Step 8: Determine Joint and Marginal Probabilities

We compute the *joints* and *marginals* for $P(D4_j, C2_i, B2_i, A_i)$ and report these in
Table 9.109.

Table 9.106 Node $B2$: likelihood probabilities

$P(B2_j|A_i)$			
$B2_j$			
A_i	True	False	Total
True	0.302900[a]	0.697100	1.000000
False	0.302900	0.697100	1.000000

[a]From Table 6.10, 0.302900 = 3029/10000

Table 9.107 Node $C2$: likelihood probabilities

$P(C2_j|B2_i, A_i)$				
$C2_j$				
$B2_i$	A_i	True	False	Total
True	True	0.481677[a]	0.518323	1.000000
True	False	0.481677	0.518323	1.000000
False	True	0.508679	0.491321	1.000000
False	False	0.508679	0.491321	1.000000

[a]From Table 6.15, 0.481677 = 1459/3029

Table 9.108 Node $D4$: likelihood probabilities

$P(D4_j|C2_i, B2_i, A_i)$					
$D4_j$					
$C2_i$	$B2_i$	A_i	True	False	Total
True	True	True	0.685811[a]	0.314189	1.000000
True	True	False	0.706915	0.293085	1.000000
True	False	True	0.748281	0.251719	1.000000
True	False	False	0.713485	0.286515	1.000000
False	True	True	0.685811	0.314189	1.000000
False	True	False	0.706915	0.293085	1.000000
False	False	True	0.748281	0.251719	1.000000
False	False	False	0.713485	0.286515	1.000000

[a]From Table 6.28, 0.685811 = 203/296

Step 9: Compute Posterior Probabilities

We compute the *posteriors* for $P(C2_j|D4_i, B2_i, A_i)$ and report this in Table 9.110.
We compute the *posteriors* for $P(B2_j|D4_i, C2_i, A_i)$ and report this in Table 9.111.
We compute the *posteriors* for $P(A_j|D4_i, C2_i, B2_i)$ and report this in Table 9.112.

9.4.15   Path 18: BBN Solution Protocol

Step 2: Specify the BBN–[$A \rightarrow D3|B2 \rightarrow C2|C2 \rightarrow D3$]

See Fig. 9.15.

Table 9.109 Joint and marginal probabilities

$P(D4_j, C2_i, B2_i, A_i)$

$D4_j$

$C2_i$	$B2_i$	A_i	True	False	Marginals
True	True	True	0.010236[a]	0.004689	0.014926
True	True	False	0.092588	0.038387	0.130974
True	False	True	0.027144	0.009131	0.036276
True	False	False	0.227120	0.091205	0.318324
False	True	True	0.011015	0.005046	0.016061
False	True	False	0.099632	0.041307	0.140939
False	False	True	0.026218	0.008820	0.035038
False	False	False	0.219370	0.088093	0.307462
Marginals			0.713322	0.286678	1.000000

Note: ZSMS $= 0.000078 - 0.000078 = 0.000000$
[a]From Tables 9.105, 9.106, 9.107, and 9.108, $0.010236 = 0.685811 * 0.481677 * 0.302900 * 0.102300$

Table 9.110 Posterior probabilities

$P(C2_j | D4_i, B2_i, A_i)$

$C2_j$

$D4_i$	$B2_i$	A_i	True	False	Total
True	True	True	0.481677[a]	0.518323	1.000000
True	True	False	0.481677	0.518323	1.000000
True	False	True	0.508679	0.491321	1.000000
True	False	False	0.508679	0.491321	1.000000
False	True	True	0.481677	0.518323	1.000000
False	True	False	0.481677	0.518323	1.000000
False	False	True	0.508679	0.491321	1.000000
False	False	False	0.508679	0.491321	1.000000

[a]From Table 9.109, $0.481677 = 0.010236/(0.010236 + 0.011015)$

Table 9.111 Posterior probabilities

$P(B2_j | D4_i, C2_i, A_i)$

$B2_j$

$D4_i$	$C2_i$	A_i	True	False	Total
True	True	True	0.273836[a]	0.726164	1.000000
True	True	False	0.289602	0.710398	1.000000
True	False	True	0.295837	0.704163	1.000000
True	False	False	0.312324	0.687676	1.000000
False	True	True	0.339306	0.660694	1.000000
False	True	False	0.296213	0.703787	1.000000
False	False	True	0.363930	0.636070	1.000000
False	False	False	0.319221	0.680779	1.000000

[a]From Table 9.109, $0.273836 = 0.010236/(0.010236 + 0.027144)$

Table 9.112 Posterior probabilities

$P(A_j\|D4_i, C2_i, B2_i)$					
A_j					
$D4_i$	$C2_i$	$B2_i$	True	False	Total
True	True	True	0.099550[a]	0.900450	1.000000
True	True	False	0.106756	0.893244	1.000000
True	False	True	0.099550	0.900450	1.000000
True	False	False	0.106756	0.893244	1.000000
False	True	True	0.108865	0.891135	1.000000
False	True	False	0.091007	0.908993	1.000000
False	False	True	0.108865	0.891135	1.000000
False	False	False	0.091007	0.908993	1.000000

[a]From Table 9.109, $0.099550 = 0.010236/(0.010236 + 0.092588)$

Fig. 9.15 Paths $[A \rightarrow D3\|B2 \rightarrow C2\|C2 \rightarrow D3]$

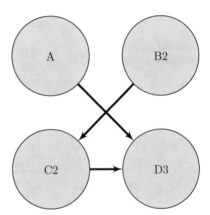

Step 6: Determine Prior or Unconditional Probabilities

We select Node A for this BBN structure as reported in Table 9.113.

Table 9.113 Node A: prior probabilities

$P(A_j)$		
True	False	Total
0.102300[a]	0.897700	1.000000

[a]From Table 6.5, $0.102300 = 1023/10000$

Step 7: Determine Likelihood Probabilities

We select Node $B2$ for this BBN structure as reported in Table 9.114.
We select Node $C2$ for this BBN structure as reported in Table 9.115.

Table 9.114 Node $B2$: likelihood probabilities

$P(B2_j|A_i)$			
$B2_j$			
A_i	True	False	Total
True	0.302900[a]	0.697100	1.000000
False	0.302900	0.697100	1.000000

[a]From Table 6.10, $0.302900 = 3029/10000$

Table 9.115 Node $C2$: likelihood probabilities

$P(C2_j|B2_i, A_i)$				
$C2_j$				
$B2_i$	A_i	True	False	Total
True	True	0.481677[a]	0.518323	1.000000
True	False	0.481677	0.518323	1.000000
False	True	0.508679	0.491321	1.000000
False	False	0.508679	0.491321	1.000000

[a]From Table 6.15, $0.481677 = 1459/3029$

We select Node $D3$ for this BBN structure as reported in Table 9.116.

Table 9.116 Node $D3$: likelihood probabilities

$P(D3_j|C2_i, B2_i, A_i)$					
$D3_j$					
$C2_i$	$B2_i$	A_i	True	False	Total
True	True	True	0.716049[a]	0.283951	1.000000
True	True	False	0.709892	0.290108	1.000000
True	False	True	0.716049	0.283951	1.000000
True	False	False	0.709892	0.290108	1.000000
False	True	True	0.743017	0.256983	1.000000
False	True	False	0.713100	0.286900	1.000000
False	False	True	0.743017	0.256983	1.000000
False	False	False	0.713100	0.286900	1.000000

[a]From Table 6.26, $0.716049 = 348/486$

Step 8: Determine Joint and Marginal Probabilities

We compute the *joints* and *marginals* for $P(D3_j, C2_i, B2_i, A_i)$ and report these in Table 9.117.

Step 9: Compute Posterior Probabilities

We compute the *posteriors* for $P(C2_j|D3_i, B2_i, A_i)$ and report this in Table 9.118.

Table 9.117 Joint and marginal probabilities

$P(D3_j, C2_i, B2_i, A_i)$

$D3_j$

$C2_i$	$B2_i$	A_i	True	False	Marginals
True	True	True	0.010687[a]	0.004238	0.014926
True	True	False	0.092978	0.037997	0.130974
True	False	True	0.025975	0.010300	0.036276
True	False	False	0.225976	0.092349	0.318324
False	True	True	0.011934	0.004127	0.016061
False	True	False	0.100504	0.040435	0.140939
False	False	True	0.026034	0.009004	0.035038
False	False	False	0.219251	0.088211	0.307462
Marginals			0.713338	0.286662	1.000000

Note: ZSMS $= 0.000062 - 0.000062 = 0.000000$
[a]From Tables 9.113, 9.114, 9.115, and 9.116, $0.010687 = 0.716049 * 0.481677 * 0.302900 * 0.102300$

Table 9.118 Posterior probabilities

$P(C2_j | D3_i, B2_i, A_i)$

$C2_j$

$D3_i$	$B2_i$	A_i	True	False	Total
True	True	True	0.472454[a]	0.527546	1.000000
True	True	False	0.480551	0.519449	1.000000
True	False	True	0.499437	0.500563	1.000000
True	False	False	0.507552	0.492448	1.000000
False	True	True	0.506616	0.493384	1.000000
False	True	False	0.484454	0.515546	1.000000
False	False	True	0.533576	0.466424	1.000000
False	False	False	0.511458	0.488542	1.000000

[a]From Table 9.117, $0.472454 = 0.010687/(0.010687 + 0.011934)$

We compute the *posteriors* for $P(B2_j | D3_i, C2_i, A_i)$ and report this in Table 9.119. We compute the *posteriors* for $P(A_j | D3_i, C2_i, B2_i)$ and report this in Table 9.120.

9.4.16   Path 19: BBN Solution Protocol

Step 2: Specify the BBN–[$A \rightarrow D1 | B2 \rightarrow D1 | C4 \rightarrow D1$]

See Fig. 9.16.

Table 9.119 Posterior probabilities

$P(B2_j$	$\vert D3_i, C2_i, A_i)$				
$B2_j$					
$D3_i$	$C2_i$	A_i	True	False	Total
True	True	True	0.291508[a]	0.708492	1.000000
True	True	False	0.291508	0.708492	1.000000
True	False	True	0.314314	0.685686	1.000000
True	False	False	0.314314	0.685686	1.000000
False	True	True	0.291508	0.708492	1.000000
False	True	False	0.291508	0.708492	1.000000
False	False	True	0.314314	0.685686	1.000000
False	False	False	0.314314	0.685686	1.000000

[a]From Table 9.117, $0.291508 = 0.010687/(0.010687 + 0.025975)$

Table 9.120 Posterior probabilities

$P(A_j$	$\vert D3_i, C2_i, B2_i)$				
A_j					
$D3_i$	$C2_i$	$B2_i$	True	False	Total
True	True	True	0.103096[a]	0.896904	1.000000
True	True	False	0.103096	0.896904	1.000000
True	False	True	0.106136	0.893864	1.000000
True	False	False	0.106136	0.893864	1.000000
False	True	True	0.100346	0.899654	1.000000
False	True	False	0.100346	0.899654	1.000000
False	False	True	0.092621	0.907379	1.000000
False	False	False	0.092621	0.907379	1.000000

[a]From Table 9.117, $0.103096 = 0.010687/(0.010687 + 0.092978)$

Fig. 9.16 Paths $[A \rightarrow D1 \vert B2 \rightarrow D1 \vert C4 \rightarrow D1]$

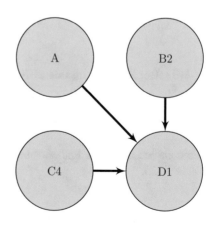

Step 6: Determine Prior or Unconditional Probabilities

We select Node A for this BBN structure as reported in Table 9.121.

Table 9.121 Node A: prior
probabilities

$P(A_j)$		
True	False	Total
0.102300[a]	0.897700	1.000000

[a]From   Table   6.5,   0.102300   =
1023/10000

Step 7: Determine Likelihood Probabilities

We select Node $B2$ for this BBN structure as reported in Table 9.122.

Table 9.122 Node $B2$:
likelihood probabilities

$P(B2_j|A_i)$			
$B2_j$			
A_i	True	False	Total
True	0.302900[a]	0.697100	1.000000
False	0.302900	0.697100	1.000000

[a]From Table 6.10, 0.302900 = 3029/10000

We select Node $C4$ for this BBN structure as reported in Table 9.123.

Table 9.123 Node $C4$:
likelihood probabilities

$P(C4_j|B2_i, A_i)$				
$C4_j$				
$B2_i$	A_i	True	False	Total
True	True	0.500500[a]	0.499500	1.000000
True	False	0.500500	0.499500	1.000000
False	True	0.500500	0.499500	1.000000
False	False	0.500500	0.499500	1.000000

[a]From Table 6.19, 0.500500 = 5005/10000

We select Node $D1$ for this BBN structure as reported in Table 9.124.

Step 8: Determine Joint and Marginal Probabilities

We compute the *joints* and *marginals* for $P(D1_j, C4_i, B2_i, A_i)$ and report these in
Table 9.125.

Step 9: Compute Posterior Probabilities

We compute the *posteriors* for $P(C4_j|D1_i, B2_i, A_i)$ and report this in Table 9.126.
We compute the *posteriors* for $P(B2_j|D1_i, C4_i, A_i)$ and report this in Table 9.127.
We compute the *posteriors* for $P(A_j|D1_i, C4_i, B2_i)$ and report this in Table 9.128.

Table 9.124 Node $D1$: likelihood probabilities

$P(D1_j$	$C4_i, B2_i, A_i)$				
$D1_j$					
$C4_i$	$B2_i$	A_i	True	False	Total
True	True	True	0.637681^a	0.362319	1.000000
True	True	False	0.698713	0.301287	1.000000
True	False	True	0.747126	0.252874	1.000000
True	False	False	0.714509	0.285491	1.000000
False	True	True	0.727848	0.272152	1.000000
False	True	False	0.714589	0.285411	1.000000
False	False	True	0.749340	0.250660	1.000000
False	False	False	0.712410	0.287590	1.000000

[a]From Table 6.22, $0.637681 = 88/138$

Table 9.125 Joint and marginal probabilities

$P(D1_j, C4_i, B2_i, A_i)$					
$D1_j$					
$C4_i$	$B2_i$	A_i	True	False	Marginals
True	True	True	0.009890^a	0.005619	0.015509
True	True	False	0.095090	0.041003	0.136093
True	False	True	0.026667	0.009026	0.035692
True	False	False	0.223789	0.089418	0.313206
False	True	True	0.011266	0.004212	0.015478
False	True	False	0.097056	0.038765	0.135821
False	False	True	0.026692	0.008929	0.035621
False	False	False	0.222685	0.089895	0.312580
Marginals			0.713134	0.286866	1.000000

Note: ZSMS $= 0.000266 - 0.000266 = 0.000000$
[a]From Tables 9.121, 9.122, 9.123, and 9.124, $0.009890 = 0.637681 * 0.500500 * 0.302900 * 0.102300$

Table 9.126 Posterior probabilities

$P(C4_j$	$D1_i, B2_i, A_i)$				
$C4_j$					
$D1_i$	$B2_i$	A_i	True	False	Total
True	True	True	0.467482^a	0.532518	1.000000
True	True	False	0.494883	0.505117	1.000000
True	False	True	0.499760	0.500240	1.000000
True	False	False	0.501236	0.498764	1.000000
False	True	True	0.571547	0.428453	1.000000
False	True	False	0.514030	0.485970	1.000000
False	False	True	0.502698	0.497302	1.000000
False	False	False	0.498668	0.501332	1.000000

[a]From Table 9.125, $0.467482 = 0.009890/(0.009890 + 0.011266)$

Table 9.127 Posterior
probabilities

$P(B2_j\|D1_i, C4_i, A_i)$					
$B2_j$					
$D1_i$	$C4_i$	A_i	True	False	Total
True	True	True	0.270533[a]	0.729467	1.000000
True	True	False	0.298201	0.701799	1.000000
True	False	True	0.296791	0.703209	1.000000
True	False	False	0.303545	0.696455	1.000000
False	True	True	0.383696	0.616304	1.000000
False	True	False	0.314390	0.685610	1.000000
False	False	True	0.320546	0.679454	1.000000
False	False	False	0.301296	0.698704	1.000000

[a]From Table 9.125, $0.270533 = 0.009890/(0.009890 + 0.026667)$

Table 9.128 Posterior
probabilities

$P(A_j\|D1_i, C4_i, B2_i)$					
A_j					
$D1_i$	$C4_i$	$B2_i$	True	False	Total
True	True	True	0.094206[a]	0.905794	1.000000
True	True	False	0.106473	0.893527	1.000000
True	False	True	0.104001	0.895999	1.000000
True	False	False	0.107036	0.892964	1.000000
False	True	True	0.120525	0.879475	1.000000
False	True	False	0.091684	0.908316	1.000000
False	False	True	0.098013	0.901987	1.000000
False	False	False	0.090350	0.909650	1.000000

[a]From Table 9.125, $0.094206 = 0.009890/(0.009890 + 0.095090)$

9.5   4-Path BBN

9.5.1   Path 1: BBN Solution Protocol

Step 2: Specify the BBN–[$A \rightarrow B1|A \rightarrow C1|A \rightarrow D7|B1 \rightarrow C1$]

See Fig. 9.17.

Step 6: Determine Prior or Unconditional Probabilities

We select Node A for this BBN structure as reported in Table 9.129.

Fig. 9.17 Paths
$[A \rightarrow B1|A \rightarrow C1|A \rightarrow D7|B1 \rightarrow C1]$

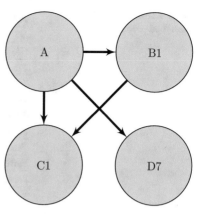

Table 9.129 Node A: prior
probabilities

$P(A_j)$		
True	False	Total
0.102300[a]	0.897700	1.000000

[a]From   Table   6.5,   0.102300 =
1023/10000

Step 7: Determine Likelihood Probabilities

We select Node $B1$ for this BBN structure as reported in Table 9.130.

Table 9.130 Node $B1$:
likelihood probabilities

$P(B1_j|A_i)$			
$B1_j$			
A_i	True	False	Total
True	0.289345[a]	0.710655	1.000000
False	0.304445	0.695555	1.000000

[a]From Table 6.8, 0.289345 = 296/1023

We select Node $C1$ for this BBN structure as reported in Table 9.131.

Table 9.131 Node $C1$:
likelihood probabilities

$P(C1_j|B1_i, A_i)$				
$C1_j$				
$B1_i$	A_i	True	False	Total
True	True	0.466216[a]	0.533784	1.000000
True	False	0.483352	0.516648	1.000000
False	True	0.478680	0.521320	1.000000
False	False	0.512172	0.487828	1.000000

[a]From Table 6.13, 0.466216 = 138/296

We select Node $D7$ for this BBN structure as reported in Table 9.132.

Table 9.132 Node $D7$: likelihood probabilities

$P(D7_j \mid C1_i, B1_i, A_i)$					
$D7_j$					
$C1_i$	$B1_i$	A_i	True	False	Total
True	True	True	0.730205[a]	0.269795	1.000000
True	True	False	0.711485	0.288515	1.000000
True	False	True	0.730205	0.269795	1.000000
True	False	False	0.711485	0.288515	1.000000
False	True	True	0.730205	0.269795	1.000000
False	True	False	0.711485	0.288515	1.000000
False	False	True	0.730205	0.269795	1.000000
False	False	False	0.711485	0.288515	1.000000

[a]From Table 6.34, $0.730205 = 747/1023$

Step 8: Determine Joint and Marginal Probabilities

We compute the *joints* and *marginals* for $P(D7_j, C1_i, B1_i, A_i)$ and report these in Table 9.133.

Table 9.133 Joint and marginal probabilities

$P(D7_j, C1_i, B1_i, A_i)$					
$D7_j$					
$C1_i$	$B1_i$	A_i	True	False	Marginals
True	True	True	0.010077[a]	0.003723	0.013800
True	True	False	0.093987	0.038113	0.132100
True	False	True	0.025411	0.009389	0.034800
True	False	False	0.227533	0.092267	0.319800
False	True	True	0.011537	0.004263	0.015800
False	True	False	0.100462	0.040738	0.141200
False	False	True	0.027675	0.010225	0.037900
False	False	False	0.216718	0.087882	0.304600
Marginals			0.713400	0.286600	1.000000

Note: ZSMS $= 0.000000 + 0.000000 = 0.000000$
[a]From Tables 9.129, 9.130, 9.131, and 9.132, $0.010077 = 0.730205 * 0.466216 * 0.289345 * 0.102300$

Step 9: Compute Posterior Probabilities

We compute the *posteriors* for $P(C1_j \mid D7_i, B1_i, A_i)$ and report this in Table 9.134.
We compute the *posteriors* for $P(B1_j \mid D7_i, C1_i, A_i)$ and report this in Table 9.135.
We compute the *posteriors* for $P(A_j \mid D7_i, C1_i, B1_i)$ and report this in Table 9.136.

Table 9.134 Posterior probabilities

$P(C1_j | D7_i, B1_i, A_i)$

			$C1_j$		
$D7_i$	$B1_i$	A_i	True	False	Total
True	True	True	0.466216[a]	0.533784	1.000000
True	True	False	0.483352	0.516648	1.000000
True	False	True	0.478680	0.521320	1.000000
True	False	False	0.512172	0.487828	1.000000
False	True	True	0.466216	0.533784	1.000000
False	True	False	0.483352	0.516648	1.000000
False	False	True	0.478680	0.521320	1.000000
False	False	False	0.512172	0.487828	1.000000

[a]From Table 9.133, $0.466216 = 0.010077/(0.010077 + 0.011537)$

Table 9.135 Posterior probabilities

$P(B1_j | D7_i, C1_i, A_i)$

			$B1_j$		
$D7_i$	$C1_i$	A_i	True	False	Total
True	True	True	0.283951[a]	0.716049	1.000000
True	True	False	0.292321	0.707679	1.000000
True	False	True	0.294227	0.705773	1.000000
True	False	False	0.316734	0.683266	1.000000
False	True	True	0.283951	0.716049	1.000000
False	True	False	0.292321	0.707679	1.000000
False	False	True	0.294227	0.705773	1.000000
False	False	False	0.316734	0.683266	1.000000

[a]From Table 9.133, $0.283951 = 0.010077/(0.010077 + 0.025411)$

Table 9.136 Posterior probabilities

$P(A_j | D7_i, C1_i, B1_i)$

			A_j		
$D7_i$	$C1_i$	$B1_i$	True	False	Total
True	True	True	0.096833[a]	0.903167	1.000000
True	True	False	0.100462	0.899538	1.000000
True	False	True	0.103012	0.896988	1.000000
True	False	False	0.113239	0.886761	1.000000
False	True	True	0.088994	0.911006	1.000000
False	True	False	0.092359	0.907641	1.000000
False	False	True	0.094726	0.905274	1.000000
False	False	False	0.104225	0.895775	1.000000

[a]From Table 9.133, $0.096833 = 0.010077/(0.010077 + 0.093987)$

Fig. 9.18 Paths
$[A \rightarrow B1|A \rightarrow C3|A \rightarrow D4|B1 \rightarrow D4]$

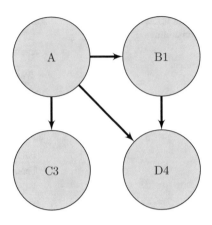

9.5.2 Path 2: BBN Solution Protocol

Step 2: Specify the BBN–[$A \rightarrow B1|A \rightarrow C3|A \rightarrow D4|B1 \rightarrow D4$]

See Fig. 9.18.

Step 6: Determine Prior or Unconditional Probabilities

We select Node A for this BBN structure as reported in Table 9.137.

Table 9.137 Node A: prior probabilities

$P(A_j)$		
True	False	Total
0.102300[a]	0.897700	1.000000

[a]From Table 6.5, $0.102300 = 1023/10000$

Step 7: Determine Likelihood Probabilities

We select Node $B1$ for this BBN structure as reported in Table 9.138.
We select Node $C3$ for this BBN structure as reported in Table 9.139.
We select Node $D4$ for this BBN structure as reported in Table 9.140.

Step 8: Determine Joint and Marginal Probabilities

We compute the *joints* and *marginals* for $P(D4_j, C3_i, B1_i, A_i)$ and report these in Table 9.141.

Table 9.138 Node $B1$:
likelihood probabilities

$P(B1_j\|A_i)$			
$B1_j$			
A_i	True	False	Total
True	0.289345[a]	0.710655	1.000000
False	0.304445	0.695555	1.000000

[a]From Table 6.8, 0.289345 = 296/1023

Table 9.139 Node $C3$:
likelihood probabilities

$P(C3_j\|B1_i, A_i)$				
$C3_j$				
$B1_i$	A_i	True	False	Total
True	True	0.475073[a]	0.524927	1.000000
True	False	0.503398	0.496602	1.000000
False	True	0.475073	0.524927	1.000000
False	False	0.503398	0.496602	1.000000

[a]From Table 6.17, 0.475073 = 486/1023

Table 9.140 Node $D4$:
likelihood probabilities

$P(D4_j\|C3_i, B1_i, A_i)$					
$D4_j$					
$C3_i$	$B1_i$	A_i	True	False	Total
True	True	True	0.685811[a]	0.314189	1.000000
True	True	False	0.706915	0.293085	1.000000
True	False	True	0.748281	0.251719	1.000000
True	False	False	0.713485	0.286515	1.000000
False	True	True	0.685811	0.314189	1.000000
False	True	False	0.706915	0.293085	1.000000
False	False	True	0.748281	0.251719	1.000000
False	False	False	0.713485	0.286515	1.000000

[a]From Table 6.28, 0.685811 = 203/296

Step 9: Compute Posterior Probabilities

We compute the *posteriors* for $P(C3_j\|D4_i, B1_i, A_i)$ and report this in Table 9.142.
We compute the *posteriors* for $P(B1_j\|D4_i, C3_i, A_i)$ and report this in Table 9.143.
We compute the *posteriors* for $P(A_j\|D4_i, C3_i, B1_i)$ and report this in Table 9.144.

9.5.3   Path 3: BBN Solution Protocol

Step 2: Specify the BBN–[$A \rightarrow B1\|A \rightarrow C3\|A \rightarrow D3\|C3 \rightarrow D3$]

See Fig. 9.19.

Table 9.141 Joint and marginal probabilities

$P(D4_j, C3_i, B1_i, A_i)$

			$D4_j$		
$C3_i$	$B1_i$	A_i	True	False	Marginals
True	True	True	0.009644[a]	0.004418	0.014062
True	True	False	0.097256	0.040322	0.137579
True	False	True	0.025844	0.008694	0.034538
True	False	False	0.224264	0.090058	0.314321
False	True	True	0.010656	0.004882	0.015538
False	True	False	0.095944	0.039778	0.135721
False	False	True	0.028556	0.009606	0.038162
False	False	False	0.221236	0.088842	0.310079
Marginals			0.713400	0.286600	1.000000

Note: ZSMS = 0.000000 + 0.000000 = 0.000000
[a]From Tables 9.137, 9.138, 9.139, and 9.140, 0.009644 = 0.685811 * 0.475073 * 0.289345 * 0.102300

Table 9.142 Posterior probabilities

$P(C3_j | D4_i, B1_i, A_i)$

			$C3_j$		
$D4_i$	$B1_i$	A_i	True	False	Total
True	True	True	0.475073[a]	0.524927	1.000000
True	True	False	0.503398	0.496602	1.000000
True	False	True	0.475073	0.524927	1.000000
True	False	False	0.503398	0.496602	1.000000
False	True	True	0.475073	0.524927	1.000000
False	True	False	0.503398	0.496602	1.000000
False	False	True	0.475073	0.524927	1.000000
False	False	False	0.503398	0.496602	1.000000

[a]From Table 9.141, 0.475073 = 0.009644/(0.009644 + 0.010656)

Table 9.143 Posterior probabilities

$P(B1_j | D4_i, C3_i, A_i)$

			$B1_j$		
$D4_i$	$C3_i$	A_i	True	False	Total
True	True	True	0.271754[a]	0.728246	1.000000
True	True	False	0.302489	0.697511	1.000000
True	False	True	0.271754	0.728246	1.000000
True	False	False	0.302489	0.697511	1.000000
False	True	True	0.336957	0.663043	1.000000
False	True	False	0.309266	0.690734	1.000000
False	False	True	0.336957	0.663043	1.000000
False	False	False	0.309266	0.690734	1.000000

[a]From Table 9.141, 0.271754 = 0.009644/(0.009644 + 0.025844)

Table 9.144 Posterior probabilities

$P(A_j|D4_i, C3_i, B1_i)$					
A_j					
$D4_i$	$C3_i$	$B1_i$	True	False	Total
True	True	True	0.090215[a]	0.909785	1.000000
True	True	False	0.103331	0.896669	1.000000
True	False	True	0.099963	0.900037	1.000000
True	False	False	0.114319	0.885681	1.000000
False	True	True	0.098752	0.901248	1.000000
False	True	False	0.088037	0.911963	1.000000
False	False	True	0.109312	0.890688	1.000000
False	False	False	0.097576	0.902424	1.000000

[a]From Table 9.141, 0.090215 = 0.009644/(0.009644 + 0.097256)

Fig. 9.19 Paths
$[A \to B1|A \to C3|A \to D3|C3 \to D3]$

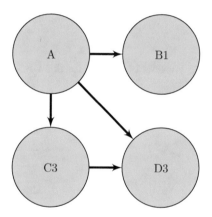

Step 6: Determine Prior or Unconditional Probabilities

We select Node A for this BBN structure as reported in Table 9.145.

Table 9.145 Node A: prior probabilities

$P(A_j)$		
True	False	Total
0.102300[a]	0.897700	1.000000

[a]From Table 6.5, 0.102300 = 1023/10000

Step 7: Determine Likelihood Probabilities

We select Node $B1$ for this BBN structure as reported in Table 9.146.
We select Node $C3$ for this BBN structure as reported in Table 9.147.

Table 9.146 Node $B1$: likelihood probabilities

$P(B1_j|A_i)$			
$B1_j$			
A_i	True	False	Total
True	0.289345[a]	0.710655	1.000000
False	0.304445	0.695555	1.000000

[a]From Table 6.8, 0.289345 = 296/1023

Table 9.147 Node $C3$: likelihood probabilities

$P(C3_j|B1_i, A_i)$					
$C3_j$					
$B1_i$	A_i	True	False	Total	
True	True	0.475073[a]	0.524927	1.000000	
True	False	0.503398	0.496602	1.000000	
False	True	0.475073	0.524927	1.000000	
False	False	0.503398	0.496602	1.000000	

[a]From Table 6.17, 0.475073 = 486/1023

We select Node $D3$ for this BBN structure as reported in Table 9.148.

Table 9.148 Node $D3$: likelihood probabilities

$P(D3_j|C3_i, B1_i, A_i)$					
$D3_j$					
$C3_i$	$B1_i$	A_i	True	False	Total
True	True	True	0.716049[a]	0.283951	1.000000
True	True	False	0.709892	0.290108	1.000000
True	False	True	0.716049	0.283951	1.000000
True	False	False	0.709892	0.290108	1.000000
False	True	True	0.743017	0.256983	1.000000
False	True	False	0.713100	0.286900	1.000000
False	False	True	0.743017	0.256983	1.000000
False	False	False	0.713100	0.286900	1.000000

[a]From Table 6.26, 0.716049 = 348/486

Step 8: Determine Joint and Marginal Probabilities

We compute the *joints* and *marginals* for $P(D3_j, C3_i, B1_i, A_i)$ and report these in Table 9.149.

Step 9: Compute Posterior Probabilities

We compute the *posteriors* for $P(C3_j|D3_i, B1_i, A_i)$ and report this in Table 9.150.

Table 9.149 Joint and marginal probabilities

$P(D3_j, C3_i, B1_i, A_i)$

			$D3_j$		
$C3_i$	$B1_i$	A_i	True	False	Marginals
True	True	True	0.010069[a]	0.003993	0.014062
True	True	False	0.097666	0.039913	0.137579
True	False	True	0.024731	0.009807	0.034538
True	False	False	0.223134	0.091187	0.314321
False	True	True	0.011545	0.003993	0.015538
False	True	False	0.096783	0.038938	0.135721
False	False	True	0.028355	0.009807	0.038162
False	False	False	0.221117	0.088962	0.310079
Marginals			0.713400	0.286600	1.000000

Note: ZSMS = 0.000000 + 0.000000 = 0.000000
[a]From Tables 9.145, 9.146, 9.147, and 9.148, 0.010069 = 0.716049 * 0.475073 * 0.289345 * 0.102300

Table 9.150 Posterior probabilities

$P(C3_j | D3_i, B1_i, A_i)$

			$C3_j$		
$D3_i$	$B1_i$	A_i	True	False	Total
True	True	True	0.465863[a]	0.534137	1.000000
True	True	False	0.502270	0.497730	1.000000
True	False	True	0.465863	0.534137	1.000000
True	False	False	0.502270	0.497730	1.000000
False	True	True	0.500000	0.500000	1.000000
False	True	False	0.506178	0.493822	1.000000
False	False	True	0.500000	0.500000	1.000000
False	False	False	0.506178	0.493822	1.000000

[a]From Table 9.149, 0.465863 = 0.010069/(0.010069 + 0.011545)

We compute the *posteriors* for $P(B1_j | D3_i, C3_i, A_i)$ and report this in Table 9.151. We compute the *posteriors* for $P(A_j | D3_i, C3_i, B1_i)$ and report this in Table 9.152.

9.5.4   Path 4: BBN Solution Protocol

Step 2: Specify the BBN–[A $\rightarrow$ B1|A $\rightarrow$ C1|B1 $\rightarrow$ C1|B1 $\rightarrow$ D6]

See Fig. 9.20.

Table 9.151 Posterior probabilities

$P(B1_j|D3_i, C3_i, A_i)$					
$B1_j$					
$D3_i$	$C3_i$	A_i	True	False	Total
True	True	True	0.289345[a]	0.710655	1.000000
True	True	False	0.304445	0.695555	1.000000
True	False	True	0.289345	0.710655	1.000000
True	False	False	0.304445	0.695555	1.000000
False	True	True	0.289345	0.710655	1.000000
False	True	False	0.304445	0.695555	1.000000
False	False	True	0.289345	0.710655	1.000000
False	False	False	0.304445	0.695555	1.000000

[a]From Table 9.149, 0.289345 = 0.010069/(0.010069 + 0.024731)

Table 9.152 Posterior probabilities

$P(A_j|D3_i, C3_i, B1_i)$					
A_j					
$D3_i$	$C3_i$	$B1_i$	True	False	Total
True	True	True	0.093463[a]	0.906537	1.000000
True	True	False	0.099775	0.900225	1.000000
True	False	True	0.106573	0.893427	1.000000
True	False	False	0.113661	0.886339	1.000000
False	True	True	0.090944	0.909056	1.000000
False	True	False	0.097105	0.902895	1.000000
False	False	True	0.093008	0.906992	1.000000
False	False	False	0.099293	0.900707	1.000000

[a]From Table 9.149, 0.093463 = 0.010069/(0.010069 + 0.097666)

Step 6: Determine Prior or Unconditional Probabilities

We select Node *A* for this BBN structure as reported in Table 9.153.

Table 9.153 Node *A*: prior probabilities

$P(A_j)$		
True	False	Total
0.102300[a]	0.897700	1.000000

[a]From Table 6.5, 0.102300 = 1023/10000

Step 7: Determine Likelihood Probabilities

We select Node *B*1 for this BBN structure as reported in Table 9.154.

Fig. 9.20 Paths
$[A \rightarrow B1 | A \rightarrow C1 | B1 \rightarrow C1 | B1 \rightarrow D6]$

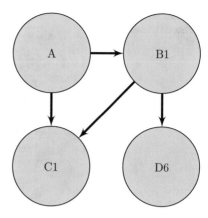

Table 9.154 Node $B1$: likelihood probabilities

$P(B1_j|A_i)$			
$B1_j$			
A_i	True	False	Total
True	0.289345[a]	0.710655	1.000000
False	0.304445	0.695555	1.000000

[a]From Table 6.8, $0.289345 = 296/1023$

We select Node $C1$ for this BBN structure as reported in Table 9.155.

Table 9.155 Node $C1$: likelihood probabilities

$P(C1_j|B1_i, A_i)$				
$C1_j$				
$B1_i$	A_i	True	False	Total
True	True	0.466216[a]	0.533784	1.000000
True	False	0.483352	0.516648	1.000000
False	True	0.478680	0.521320	1.000000
False	False	0.512172	0.487828	1.000000

[a]From Table 6.13, $0.466216 = 138/296$

We select Node $D6$ for this BBN structure as reported in Table 9.156.

Step 8: Determine Joint and Marginal Probabilities

We compute the *joints* and *marginals* for $P(D6_j, C1_i, B1_i, A_i)$ and report these in Table 9.157.

Table 9.156 Node $D6$:
likelihood probabilities

$P(D6_j | C1_i, B1_i, A_i)$

$C1_i$	$B1_i$	A_i	True	False	Total
True	True	True	0.704853[a]	0.295147	1.000000
True	True	False	0.704853	0.295147	1.000000
True	False	True	0.717114	0.282886	1.000000
True	False	False	0.717114	0.282886	1.000000
False	True	True	0.704853	0.295147	1.000000
False	True	False	0.704853	0.295147	1.000000
False	False	True	0.717114	0.282886	1.000000
False	False	False	0.717114	0.282886	1.000000

[a]From Table 6.32, $0.704853 = 2135/3029$

Table 9.157 Joint and
marginal probabilities

$P(D6_j, C1_i, B1_i, A_i)$

$D6_j$

$C1_i$	$B1_i$	A_i	True	False	Marginals
True	True	True	0.009727[a]	0.004073	0.013800
True	True	False	0.093111	0.038989	0.132100
True	False	True	0.024956	0.009844	0.034800
True	False	False	0.229333	0.090467	0.319800
False	True	True	0.011137	0.004663	0.015800
False	True	False	0.099525	0.041675	0.141200
False	False	True	0.027179	0.010721	0.037900
False	False	False	0.218433	0.086167	0.304600
Marginals			0.713400	0.286600	1.000000

Note: ZSMS $= 0.000000 + 0.000000 = 0.000000$
[a]From Tables 9.153, 9.154, 9.155, and 9.156, $0.009727 = 0.704853 * 0.466216 * 0.289345 * 0.102300$

Step 9: Compute Posterior Probabilities

We compute the *posteriors* for $P(C1_j | D6_i, B1_i, A_i)$ and report this in Table 9.158.
We compute the *posteriors* for $P(B1_j | D4_i, C2_i, A_i)$ and report this in Table 9.159.
We compute the *posteriors* for $P(A_j | D6_i, C1_i, B1_i)$ and report this in Table 9.160.

9.5.5 Path 5: BBN Solution Protocol

Step 2: Specify the BBN–[$A \rightarrow B1 | A \rightarrow C1 | B1 \rightarrow C1 | C1 \rightarrow D5$]

See Fig. 9.21.

Table 9.158 Posterior probabilities

$P(C1_j | D6_i, B1_i, A_i)$

			$C1_j$		
$D6_i$	$B1_i$	A_i	True	False	Total
True	True	True	0.466216[a]	0.533784	1.000000
True	True	False	0.483352	0.516648	1.000000
True	False	True	0.478680	0.521320	1.000000
True	False	False	0.512172	0.487828	1.000000
False	True	True	0.466216	0.533784	1.000000
False	True	False	0.483352	0.516648	1.000000
False	False	True	0.478680	0.521320	1.000000
False	False	False	0.512172	0.487828	1.000000

[a]From Table 9.157, 0.466216 = 0.009727/(0.009727 + 0.011137)

Table 9.159 Posterior probabilities

$P(B1_j | D6_i, C1_i, A_i)$

			$B1_j$		
$D6_i$	$C1_i$	A_i	True	False	Total
True	True	True	0.280457[a]	0.719543	1.000000
True	True	False	0.288767	0.711233	1.000000
True	False	True	0.290659	0.709341	1.000000
True	False	False	0.313014	0.686986	1.000000
False	True	True	0.292656	0.707344	1.000000
False	True	False	0.301175	0.698825	1.000000
False	False	True	0.303114	0.696886	1.000000
False	False	False	0.325987	0.674013	1.000000

[a]From Table 9.157, 0.280457 = 0.009727/(0.009727 + 0.024956)

Table 9.160 Posterior probabilities

$P(A_j | D6_i, C1_i, B1_i)$

			A_j		
$D6_i$	$C1_i$	$B1_i$	True	False	Total
True	True	True	0.094585[a]	0.905415	1.000000
True	True	False	0.098139	0.901861	1.000000
True	False	True	0.100637	0.899363	1.000000
True	False	False	0.110657	0.889343	1.000000
False	True	True	0.094585	0.905415	1.000000
False	True	False	0.098139	0.901861	1.000000
False	False	True	0.100637	0.899363	1.000000
False	False	False	0.110657	0.889343	1.000000

[a]From Table 9.157, 0.094585 = 0.009727/(0.009727 + 0.093111)

Step 6: Determine Prior or Unconditional Probabilities

We select Node A for this BBN structure as reported in Table 9.161.

Fig. 9.21 Paths
$[A \rightarrow B1 | A \rightarrow C1 | B1 \rightarrow C1 | C1 \rightarrow D5]$

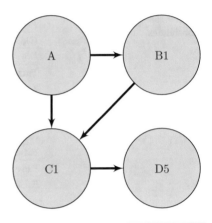

Table 9.161 Node A: prior probabilities

$P(A_j)$		
True	False	Total
0.102300[a]	0.897700	1.000000

[a]From Table 6.5, 0.102300 = 1023/10000

Step 7: Determine Likelihood Probabilities

We select Node $B1$ for this BBN structure as reported in Table 9.162.

Table 9.162 Node $B1$: likelihood probabilities

$P(B1_j | A_i)$			
$B1_j$			
A_i	True	False	Total
True	0.289345[a]	0.710655	1.000000
False	0.304445	0.695555	1.000000

[a]From Table 6.8, 0.289345 = 296/1023

We select Node $C1$ for this BBN structure as reported in Table 9.163.

Table 9.163 Node $C1$: likelihood probabilities

$P(C1_j | B1_i, A_i)$				
$C1_j$				
$B1_i$	A_i	True	False	Total
True	True	0.466216[a]	0.533784	1.000000
True	False	0.483352	0.516648	1.000000
False	True	0.478680	0.521320	1.000000
False	False	0.512172	0.487828	1.000000

[a]From Table 6.13, 0.466216 = 138/296

We select Node $D5$ for this BBN structure as reported in Table 9.164.

Table 9.164 Node $D5$: likelihood probabilities

$P(D5_j\|C1_i, \cancel{B1_i}, \cancel{A_i})$					
$D5_j$					
$C1_i$	$\cancel{B1_i}$	$\cancel{A_i}$	True	False	Total
True	True	True	0.710490[a]	0.289510	1.000000
True	True	False	0.710490	0.289510	1.000000
True	False	True	0.710490	0.289510	1.000000
True	False	False	0.710490	0.289510	1.000000
False	True	True	0.716316	0.283684	1.000000
False	True	False	0.716316	0.283684	1.000000
False	False	True	0.716316	0.283684	1.000000
False	False	False	0.716316	0.283684	1.000000

[a]From Table 6.30, $0.710490 = 3556/5005$

Step 8: Determine Joint and Marginal Probabilities

We compute the *joints* and *marginals* for $P(D5_j, C1_i, B1_i, A_i)$ and report these in Table 9.165.

Table 9.165 Joint and marginal probabilities

$P(D5_j, C1_i, B1_i, A_i)$					
$D5_j$					
$C1_i$	$B1_i$	A_i	True	False	Marginals
True	True	True	0.009805[a]	0.003995	0.013800
True	True	False	0.093856	0.038244	0.132100
True	False	True	0.024725	0.010075	0.034800
True	False	False	0.227215	0.092585	0.319800
False	True	True	0.011318	0.004482	0.015800
False	True	False	0.101144	0.040056	0.141200
False	False	True	0.027148	0.010752	0.037900
False	False	False	0.218190	0.086410	0.304600
Marginals			0.713400	0.286600	1.000000

Note: ZSMS $= 0.000000 + 0.000000 = 0.000000$
[a]From Tables 9.161, 9.162, 9.163, and 9.164, $0.009805 = 0.710490 * 0.466216 * 0.289345 * 0.102300$

Step 9: Compute Posterior Probabilities

We compute the *posteriors* for $P(C1_j|D5_i, B1_i, A_i)$ and report this in Table 9.166.
We compute the *posteriors* for $P(B1_j|D5_i, C1_i, A_i)$ and report this in Table 9.167.
We compute the *posteriors* for $P(A_j|D5_i, C1_i, B1_i)$ and report this in Table 9.168.

Table 9.166 Posterior probabilities

$P(C1_j|D5_i, B1_i, A_i)$

$C1_j$					
$D5_i$	$B1_i$	A_i	True	False	Total
True	True	True	0.464184[a]	0.535816	1.000000
True	True	False	0.481312	0.518688	1.000000
True	False	True	0.476642	0.523358	1.000000
True	False	False	0.510131	0.489869	1.000000
False	True	True	0.471279	0.528721	1.000000
False	True	False	0.488430	0.511570	1.000000
False	False	True	0.483755	0.516245	1.000000
False	False	False	0.517250	0.482750	1.000000

[a]From Table 9.165, 0.464184 = 0.009805/(0.009805 + 0.011318)

Table 9.167 Posterior probabilities

$P(B1_j|D5_i, C1_i, A_i)$

$B1_j$					
$D5_i$	$C1_i$	A_i	True	False	Total
True	True	True	0.283951[a]	0.716049	1.000000
True	True	False	0.292321	0.707679	1.000000
True	False	True	0.294227	0.705773	1.000000
True	False	False	0.316734	0.683266	1.000000
False	True	True	0.283951	0.716049	1.000000
False	True	False	0.292321	0.707679	1.000000
False	False	True	0.294227	0.705773	1.000000
False	False	False	0.316734	0.683266	1.000000

[a]From Table 9.165, 0.283951 = 0.009805/(0.009805 + 0.024725)

Table 9.168 Posterior probabilities

$P(A_j|D5_i, C1_i, B1_i)$

A_j					
$D5_i$	$C1_i$	$B1_i$	True	False	Total
True	True	True	0.094585[a]	0.905415	1.000000
True	True	False	0.098139	0.901861	1.000000
True	False	True	0.100637	0.899363	1.000000
True	False	False	0.110657	0.889343	1.000000
False	True	True	0.094585	0.905415	1.000000
False	True	False	0.098139	0.901861	1.000000
False	False	True	0.100637	0.899363	1.000000
False	False	False	0.110657	0.889343	1.000000

[a]From Table 9.165, 0.094585 = 0.009805/(0.009805 + 0.093856)

Fig. 9.22 Paths
$[A \rightarrow B1|A \rightarrow C3|B1 \rightarrow D2|C3 \rightarrow D2]$

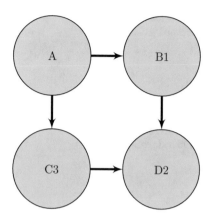

9.5.6   Path 6: BBN Solution Protocol

Step 2: Specify the BBN–$[A \rightarrow B1|A \rightarrow C3|B1 \rightarrow D2|C3 \rightarrow D2]$

See Fig. 9.22.

Step 6: Determine Prior or Unconditional Probabilities

We select Node A for this BBN structure as reported in Table 9.169.

Table 9.169 Node A: prior probabilities

$P(A_j)$		
True	False	Total
0.102300[a]	0.897700	1.000000

[a]From Table 6.5, $0.102300 = 1023/10000$

Step 7: Determine Likelihood Probabilities

We select Node $B1$ for this BBN structure as reported in Table 9.170.
We select Node $C3$ for this BBN structure as reported in Table 9.171.
We select Node $D2$ for this BBN structure as reported in Table 9.172.

Step 8: Determine Joint and Marginal Probabilities

We compute the *joints* and *marginals* for $P(D2_j, C3_i, B1_i, A_i)$ and report these in Table 9.173.

Table 9.170 Node $B1$:
likelihood probabilities

$P(B1_j|A_i)$			
$B1_j$			
A_i	True	False	Total
True	0.289345[a]	0.710655	1.000000
False	0.304445	0.695555	1.000000

[a]From Table 6.8, 0.289345 = 296/1023

Table 9.171 Node $C3$:
likelihood probabilities

$P(C3_j|B1_i, A_i)$				
$C3_j$				
$B1_i$	A_i	True	False	Total
~~True~~	True	0.475073[a]	0.524927	1.000000
~~True~~	False	0.503398	0.496602	1.000000
~~False~~	~~True~~	~~0.475073~~	~~0.524927~~	~~1.000000~~
~~False~~	~~False~~	~~0.503398~~	~~0.496602~~	~~1.000000~~

[a]From Table 6.17, 0.475073 = 486/1023

Table 9.172 Node $D2$:
likelihood probabilities

$P(D2_j|C3_i, B1_i, A_i)$					
$D2_j$					
$C3_i$	$B1_i$	A_i	True	False	Total
True	True	~~True~~	0.692940[a]	0.307060	1.000000
~~True~~	~~True~~	~~False~~	~~0.692940~~	~~0.307060~~	~~1.000000~~
True	False	~~True~~	0.717710	0.282290	1.000000
~~True~~	~~False~~	~~False~~	~~0.717710~~	~~0.282290~~	~~1.000000~~
False	True	~~True~~	0.715924	0.284076	1.000000
~~False~~	~~True~~	~~False~~	~~0.715924~~	~~0.284076~~	~~1.000000~~
False	False	~~True~~	0.716496	0.283504	1.000000
~~False~~	~~False~~	~~False~~	~~0.716496~~	~~0.283504~~	~~1.000000~~

[a]From Table 6.24, 0.692940 = 1011/1459

Step 9: Compute Posterior Probabilities

We compute the *posteriors* for $P(C3_j|D2_i, B1_i, A_i)$ and report this in Table 9.174.
We compute the *posteriors* for $P(B1_j|D2_i, C3_i, A_i)$ and report this in Table 9.175.
We compute the *posteriors* for $P(A_j|D2_i, C3_i, B1_i)$ and report this in Table 9.176.

9.5.7 Path 7: BBN Solution Protocol

Step 2: Specify the BBN–[$A \rightarrow B1|A \rightarrow D4|B1 \rightarrow C2|B1 \rightarrow D4$]

See Fig. 9.23.

Table 9.173 Joint and marginal probabilities

$P(D2_j, C3_i, B1_i, A_i)$					
$D2_j$					
$C3_i$	$B1_i$	A_i	True	False	Marginals
True	True	True	0.009744[a]	0.004318	0.014062
True	True	False	0.095334	0.042245	0.137579
True	False	True	0.024788	0.009750	0.034538
True	False	False	0.225592	0.088730	0.314321
False	True	True	0.011124	0.004414	0.015538
False	True	False	0.097166	0.038555	0.135721
False	False	True	0.027343	0.010819	0.038162
False	False	False	0.222170	0.087908	0.310079
Marginals			0.713261	0.286739	1.000000

Note: ZSMS = 0.000139 − 0.000139 = 0.000000
[a]From Tables 9.169, 9.170, 9.171, and 9.172, 0.009744 = 0.692940 * 0.475073 * 0.289345 * 0.102300

Table 9.174 Posterior probabilities

$P(C3_j|D2_i, B1_i, A_i)$					
$C3_j$					
$D2_i$	$B1_i$	A_i	True	False	Total
True	True	True	0.466944[a]	0.533056	1.000000
True	True	False	0.495240	0.504760	1.000000
True	False	True	0.475495	0.524505	1.000000
True	False	False	0.503821	0.496179	1.000000
False	True	True	0.494503	0.505497	1.000000
False	True	False	0.522831	0.477169	1.000000
False	False	True	0.474003	0.525997	1.000000
False	False	False	0.502325	0.497675	1.000000

[a]From Table 9.173, 0.466944 = 0.009744/(0.009744 + 0.011124)

Table 9.175 Posterior probabilities

$P(B1_j|D2_i, C3_i, A_i)$					
$B1_j$					
$D2_i$	$C3_i$	A_i	True	False	Total
True	True	True	0.282177[a]	0.717823	1.000000
True	True	False	0.297059	0.702941	1.000000
True	False	True	0.289181	0.710819	1.000000
True	False	False	0.304275	0.695725	1.000000
False	True	True	0.306941	0.693059	1.000000
False	True	False	0.322542	0.677458	1.000000
False	False	True	0.289760	0.710240	1.000000
False	False	False	0.304872	0.695128	1.000000

[a]From Table 9.173, 0.282177 = 0.009744/(0.009744 + 0.024788)

Table 9.176 Posterior probabilities

$P(A_j|D2_i, C3_i, B1_i)$					
A_j					
$D2_i$	$C3_i$	$B1_i$	True	False	Total
True	True	True	0.092733[a]	0.907267	1.000000
True	True	False	0.099002	0.900998	1.000000
True	False	True	0.102723	0.897277	1.000000
True	False	False	0.109586	0.890414	1.000000
False	True	True	0.092733	0.907267	1.000000
False	True	False	0.099002	0.900998	1.000000
False	False	True	0.102723	0.897277	1.000000
False	False	False	0.109586	0.890414	1.000000

[a]From Table 9.173, $0.092733 = 0.009744/(0.009744 + 0.095334)$

Fig. 9.23 Paths
$[A \rightarrow B1|A \rightarrow D4|B1 \rightarrow C2|B1 \rightarrow D4]$

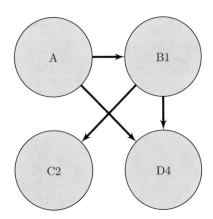

Step 6: Determine Prior or Unconditional Probabilities

We select Node A for this BBN structure as reported in Table 9.177.

Table 9.177 Node A: prior probabilities

$P(A_j)$		
True	False	Total
0.102300[a]	0.897700	1.000000

[a]From Table 6.5, $0.102300 = 1023/10000$

Step 7: Determine Likelihood Probabilities

We select Node $B1$ for this BBN structure as reported in Table 9.178.
We select Node $C2$ for this BBN structure as reported in Table 9.179.

Table 9.178 Node $B1$:
likelihood probabilities

$P(B1_j\|A_i)$			
$B1_j$			
A_i	True	False	Total
True	0.289345[a]	0.710655	1.000000
False	0.304445	0.695555	1.000000

[a]From Table 6.8, $0.289345 = 296/1023$

Table 9.179 Node $C2$:
likelihood probabilities

$P(C2_j\|B1_i, A_i)$				
$C2_j$				
$B1_i$	A_i	True	False	Total
True	True	0.481677[a]	0.518323	1.000000
True	False	0.481677	0.518323	1.000000
False	True	0.508679	0.491321	1.000000
False	False	0.508679	0.491321	1.000000

[a]From Table 6.15, $0.481677 = 1459/3029$

We select Node $D4$ for this BBN structure as reported in Table 9.180.

Table 9.180 Node $D4$:
likelihood probabilities

$P(D4_j\|C2_i, B1_i, A_i)$					
$D4_j$					
$C2_i$	$B1_i$	A_i	True	False	Total
True	True	True	0.685811[a]	0.314189	1.000000
True	True	False	0.706915	0.293085	1.000000
True	False	True	0.748281	0.251719	1.000000
True	False	False	0.713485	0.286515	1.000000
False	True	True	0.685811	0.314189	1.000000
False	True	False	0.706915	0.293085	1.000000
False	False	True	0.748281	0.251719	1.000000
False	False	False	0.713485	0.286515	1.000000

[a]From Table 6.28, $0.685811 = 203/296$

Step 8: Determine Joint and Marginal Probabilities

We compute the *joints* and *marginals* for $P(D4_j, C2_i, B1_i, A_i)$ and report these in Table 9.181.

Step 9: Compute Posterior Probabilities

We compute the *posteriors* for $P(C2_j|D4_i, B1_i, A_i)$ and report this in Table 9.182.

Table 9.181 Joint and
marginal probabilities

$P(D4_j, C2_i, B1_i, A_i)$

$D4_j$					
$C2_i$	$B1_i$	A_i	True	False	Marginals
True	True	True	0.009778[a]	0.004480	0.014258
True	True	False	0.093060	0.038582	0.131642
True	False	True	0.027672	0.009309	0.036981
True	False	False	0.226616	0.091003	0.317619
False	True	True	0.010522	0.004820	0.015342
False	True	False	0.100140	0.041518	0.141658
False	False	True	0.026728	0.008991	0.035719
False	False	False	0.218884	0.087897	0.306781
Marginals			0.713400	0.286600	1.000000

Note: ZSMS = 0.000000 + 0.000000 = 0.000000
[a]From Tables 9.177, 9.178, 9.179, and 9.180, 0.009778 = 0.685811 * 0.481677 * 0.289345 * 0.102300

Table 9.182 Posterior
probabilities

$P(C2_j | D4_i, B1_i, A_i)$

$C2_j$					
$D4_i$	$B1_i$	A_i	True	False	Total
True	True	True	0.481677[a]	0.518323	1.000000
True	True	False	0.481677	0.518323	1.000000
True	False	True	0.508679	0.491321	1.000000
True	False	False	0.508679	0.491321	1.000000
False	True	True	0.481677	0.518323	1.000000
False	True	False	0.481677	0.518323	1.000000
False	False	True	0.508679	0.491321	1.000000
False	False	False	0.508679	0.491321	1.000000

[a]From Table 9.181, 0.481677 = 0.009778/(0.009778 + 0.010522)

We compute the *posteriors* for $P(B1_j | D4_i, C2_i, A_i)$ and report this in Table 9.183. We compute the *posteriors* for $P(A_j | D4_i, C2_i, B1_i)$ and report this in Table 9.184.

9.5.8   Path 8: BBN Solution Protocol

Step 2: Specify the BBN–$[A \rightarrow B1 | A \rightarrow D3 | B1 \rightarrow C2 | C2 \rightarrow D3]$

See Fig. 9.24.

Table 9.183 Posterior probabilities

$P(B1_j | D4_i, C2_i, A_i)$

| | | | $B1_j$ | | |
$D4_i$	$C2_i$	A_i	True	False	Total
True	True	True	0.261095[a]	0.738905	1.000000
True	True	False	0.291107	0.708893	1.000000
True	False	True	0.282470	0.717530	1.000000
True	False	False	0.313895	0.686105	1.000000
False	True	True	0.324881	0.675119	1.000000
False	True	False	0.297738	0.702262	1.000000
False	False	True	0.349012	0.650988	1.000000
False	False	False	0.320810	0.679190	1.000000

[a]From Table 9.181, $0.261095 = 0.009778/(0.009778 + 0.027672)$

Table 9.184 Posterior probabilities

$P(A_j | D4_i, C2_i, B1_i)$

| | | | A_j | | |
$D4_i$	$C2_i$	$B1_i$	True	False	Total
True	True	True	0.095082[a]	0.904918	1.000000
True	True	False	0.108822	0.891178	1.000000
True	False	True	0.095082	0.904918	1.000000
True	False	False	0.108822	0.891178	1.000000
False	True	True	0.104027	0.895973	1.000000
False	True	False	0.092799	0.907201	1.000000
False	False	True	0.104027	0.895973	1.000000
False	False	False	0.092799	0.907201	1.000000

[a]From Table 9.181, $0.095082 = 0.009778/(0.009778 + 0.093060)$

Step 6: Determine Prior or Unconditional Probabilities

We select Node A for this BBN structure as reported in Table 9.185.

Table 9.185 Node A: prior probabilities

$P(A_j)$

True	False	Total
0.102300[a]	0.897700	1.000000

[a]From Table 6.5, $0.102300 = 1023/10000$

Step 7: Determine Likelihood Probabilities

We select Node $B1$ for this BBN structure as reported in Table 9.186.

Fig. 9.24 Paths
$[A \rightarrow B1|A \rightarrow D3|B1 \rightarrow C2|C2 \rightarrow D3]$

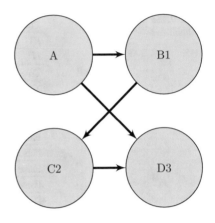

Table 9.186 Node $B1$:
likelihood probabilities

$P(B1_j|A_i)$			
$B1_j$			
A_i	True	False	Total
True	0.289345[a]	0.710655	1.000000
False	0.304445	0.695555	1.000000

[a]From Table 6.8, $0.289345 = 296/1023$

We select Node $C2$ for this BBN structure as reported in Table 9.187.

Table 9.187 Node $C2$:
likelihood probabilities

$P(C2_j|B1_i, A_i)$				
$C2_j$				
$B1_i$	A_i	True	False	Total
True	True	0.481677[a]	0.518323	1.000000
True	False	0.481677	0.518323	1.000000
False	True	0.508679	0.491321	1.000000
False	False	0.508679	0.491321	1.000000

[a]From Table 6.15, $0.481677 = 1459/3029$

We select Node $D3$ for this BBN structure as reported in Table 9.188.

Step 8: Determine Joint and Marginal Probabilities

We compute the *joints* and *marginals* for $P(D3_j, C2_i, B1_i, A_i)$ and report these in Table 9.189.

Table 9.188 Node $D3$:
likelihood probabilities

$P(D3_j$	$C2_i, B1_i, A_i)$				
$D3_j$					
$C2_i$	$B1_i$	A_i	True	False	Total
True	True	True	0.716049[a]	0.283951	1.000000
True	True	False	0.709892	0.290108	1.000000
True	False	True	0.716049	0.283951	1.000000
True	False	False	0.709892	0.290108	1.000000
False	True	True	0.743017	0.256983	1.000000
False	True	False	0.713100	0.286900	1.000000
False	False	True	0.743017	0.256983	1.000000
False	False	False	0.713100	0.286900	1.000000

[a]From Table 6.26, $0.716049 = 348/486$

Table 9.189 Joint and
marginal probabilities

$P(D3_j, C2_i, B1_i, A_i)$					
$D3_j$					
$C2_i$	$B1_i$	A_i	True	False	Marginals
True	True	True	0.010209[a]	0.004048	0.014258
True	True	False	0.093452	0.038191	0.131642
True	False	True	0.026480	0.010501	0.036981
True	False	False	0.225475	0.092144	0.317619
False	True	True	0.011400	0.003943	0.015342
False	True	False	0.101016	0.040642	0.141658
False	False	True	0.026540	0.009179	0.035719
False	False	False	0.218766	0.088015	0.306781
Marginals			0.713337	0.286663	1.000000

Note: ZSMS $= 0.000063 - 0.000063 = 0.000000$
[a]From Tables 9.185, 9.186, 9.187, and 9.188, $0.010209 = 0.716049 * 0.481677 * 0.289345 * 0.102300$

Step 9: Compute Posterior Probabilities

We compute the *posteriors* for $P(C2_j|D3_i, B1_i, A_i)$ and report this in Table 9.190.
We compute the *posteriors* for $P(B1_j|D3_i, C2_i, A_i)$ and report this in Table 9.191.
We compute the *posteriors* for $P(A_j|D3_i, C2_i, B1_i)$ and report this in Table 9.192.

9.5.9 Path 9: BBN Solution Protocol

Step 2: Specify the BBN–[$A \rightarrow B1|A \rightarrow D1|B1 \rightarrow D1|C4 \rightarrow D1$]

See Fig. 9.25.

Table 9.190 Posterior probabilities

$P(C2_j|D3_i, B1_i, A_i)$					
$C2_j$					
$D3_i$	$B1_i$	A_i	True	False	Total
True	True	True	0.472454[a]	0.527546	1.000000
True	True	False	0.480551	0.519449	1.000000
True	False	True	0.499437	0.500563	1.000000
True	False	False	0.507552	0.492448	1.000000
False	True	True	0.506616	0.493384	1.000000
False	True	False	0.484454	0.515546	1.000000
False	False	True	0.533576	0.466424	1.000000
False	False	False	0.511458	0.488542	1.000000

[a]From Table 9.189, $0.472454 = 0.010209/(0.010209 + 0.011400)$

Table 9.191 Posterior probabilities

$P(B1_j|D3_i, C2_i, A_i)$					
$B1_j$					
$D3_i$	$C2_i$	A_i	True	False	Total
True	True	True	0.278260[a]	0.721740	1.000000
True	True	False	0.293020	0.706980	1.000000
True	False	True	0.300469	0.699531	1.000000
True	False	False	0.315891	0.684109	1.000000
False	True	True	0.278260	0.721740	1.000000
False	True	False	0.293020	0.706980	1.000000
False	False	True	0.300469	0.699531	1.000000
False	False	False	0.315891	0.684109	1.000000

[a]From Table 9.189, $0.278260 = 0.010209/(0.010209 + 0.026480)$

Table 9.192 Posterior probabilities

$P(A_j|D3_i, C2_i, B1_i)$					
A_j					
$D3_i$	$C2_i$	$B1_i$	True	False	Total
True	True	True	0.098486[a]	0.901514	1.000000
True	True	False	0.105099	0.894901	1.000000
True	False	True	0.101406	0.898594	1.000000
True	False	False	0.108191	0.891809	1.000000
False	True	True	0.095847	0.904153	1.000000
False	True	False	0.102302	0.897698	1.000000
False	False	True	0.088433	0.911567	1.000000
False	False	False	0.094441	0.905559	1.000000

[a]From Table 9.189, $0.098486 = 0.010209/(0.010209 + 0.093452)$

Fig. 9.25 Paths
$[A \rightarrow B1|A \rightarrow D1|B1 \rightarrow D1|C4 \rightarrow D1]$

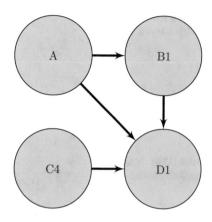

Step 6: Determine Prior or Unconditional Probabilities

We select Node *A* for this BBN structure as reported in Table 9.193.

Table 9.193 Node *A*: Prior probabilities

$P(A_j)$		
True	False	Total
0.102300[a]	0.897700	1.000000

[a]From Table 6.5, 0.102300 = 1023/10000

Step 7: Determine Likelihood Probabilities

We select Node *B*1 for this BBN structure as reported in Table 9.194.

Table 9.194 Node *B*1: likelihood probabilities

$P(B1_j|A_i)$			
$B1_j$			
A_i	True	False	Total
True	0.289345[a]	0.710655	1.000000
False	0.304445	0.695555	1.000000

[a]From Table 6.8, 0.289345 = 296/1023

We select Node *C*4 for this BBN structure as reported in Table 9.195.
We select Node *D*1 for this BBN structure as reported in Table 9.196.

Table 9.195 Node $C4$:
likelihood probabilities

$P(C4_j|B1_i, A_i)$

$C4_j$				
$B1_i$	A_i	True	False	Total
~~True~~	~~True~~	0.500500[a]	0.499500	1.000000
~~True~~	~~False~~	~~0.500500~~	~~0.499500~~	~~1.000000~~
~~False~~	~~True~~	~~0.500500~~	0.499500	~~1.000000~~
~~False~~	~~False~~	~~0.500500~~	0.499500	~~1.000000~~

[a]From Table 6.19, $0.500500 = 5005/10000$

Table 9.196 Node $D1$:
likelihood probabilities

$P(D1_j|C4_i, B1_i, A_i)$

$D1_j$					
$C4_i$	$B1_i$	A_i	True	False	Total
True	True	True	0.637681[a]	0.362319	1.000000
True	True	False	0.698713	0.301287	1.000000
True	False	True	0.747126	0.252874	1.000000
True	False	False	0.714509	0.285491	1.000000
False	True	True	0.727848	0.272152	1.000000
False	True	False	0.714589	0.285411	1.000000
False	False	True	0.749340	0.250660	1.000000
False	False	False	0.712410	0.287590	1.000000

[a]From Table 6.22, $0.637681 = 88/138$

Step 8: Determine Joint and Marginal Probabilities

We compute the *joints* and *marginals* for $P(D1_j, C4_i, B1_i, A_i)$ and report these in Table 9.197.

Table 9.197 Joint and
marginal probabilities

$P(D1_j, C4_i, B1_i, A_i)$

$D1_j$					
$C4_i$	$B1_i$	A_i	True	False	Marginals
True	True	True	0.009447[a]	0.005368	0.014815
True	True	False	0.095575	0.041212	0.136787
True	False	True	0.027185	0.009201	0.036386
True	False	False	0.223293	0.089219	0.312512
False	True	True	0.010761	0.004024	0.014785
False	True	False	0.097551	0.038962	0.136513
False	False	True	0.027211	0.009102	0.036314
False	False	False	0.222192	0.089696	0.311888
Marginals			0.713215	0.286785	1.000000

Note: ZSMS $= 0.000185 - 0.000185 = 0.000000$
[a]From Tables 9.193, 9.194, 9.195, and 9.196, $0.009447 = 0.637681 * 0.500500 * 0.289345 * 0.102300$

Step 9: Compute Posterior Probabilities

We compute the *posteriors* for $P(C4_j|D1_i, B1_i, A_i)$ and report this in Table 9.198.

Table 9.198 Posterior probabilities

$P(C4_j|D1_i, B1_i, A_i)$

			$C4_j$		
$D1_i$	$B1_i$	A_i	True	False	Total
True	True	True	0.467482[a]	0.532518	1.000000
True	True	False	0.494883	0.505117	1.000000
True	False	True	0.499760	0.500240	1.000000
True	False	False	0.501236	0.498764	1.000000
False	True	True	0.571547	0.428453	1.000000
False	True	False	0.514030	0.485970	1.000000
False	False	True	0.502698	0.497302	1.000000
False	False	False	0.498668	0.501332	1.000000

[a]From Table 9.197, $0.467482 = 0.009447/(0.009447 + 0.010761)$

We compute the *posteriors* for $P(B1_j|D1_i, C4_i, A_i)$ and report this in Table 9.199.

Table 9.199 Posterior probabilities

$P(B1_j|D1_i, C4_i, A_i)$

			$B1_j$		
$D1_i$	$C4_i$	A_i	True	False	Total
True	True	True	0.257890[a]	0.742110	1.000000
True	True	False	0.299732	0.700268	1.000000
True	False	True	0.283398	0.716602	1.000000
True	False	False	0.305092	0.694908	1.000000
False	True	True	0.368436	0.631564	1.000000
False	True	False	0.315967	0.684033	1.000000
False	False	True	0.306549	0.693451	1.000000
False	False	False	0.302836	0.697164	1.000000

[a]From Table 9.197, $0.257890 = 0.009447/(0.009447 + 0.027185)$

We compute the *posteriors* for $P(A_j|D1_i, C4_i, B1_i)$ and report this in Table 9.200.

9.5.10 Path 10: BBN Solution Protocol

Step 2: Specify the BBN–$[A \rightarrow B1|B1 \rightarrow C2|B1 \rightarrow D2|C2 \rightarrow D2]$

See Fig. 9.26.

Table 9.200 Posterior
probabilities

$P(A_j|D1_i, C4_i, B1_i)$					
A_j					
$D1_i$	$C4_i$	$B1_i$	True	False	Total
True	True	True	0.089954[a]	0.910046	1.000000
True	True	False	0.108533	0.891467	1.000000
True	False	True	0.099355	0.900645	1.000000
True	False	False	0.109106	0.890894	1.000000
False	True	True	0.115236	0.884764	1.000000
False	True	False	0.093488	0.906512	1.000000
False	False	True	0.093607	0.906393	1.000000
False	False	False	0.092131	0.907869	1.000000

[a]From Table 9.197, $0.089954 = 0.009447/(0.009447 + 0.095575)$

Fig. 9.26 Paths
$[A \rightarrow B1|B1 \rightarrow C2|B1 \rightarrow D2|C2 \rightarrow D2]$

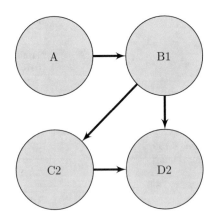

Step 6: Determine Prior or Unconditional Probabilities

We select Node A for this BBN structure as reported in Table 9.201.

Table 9.201 Node A: prior
probabilities

$P(A_j)$		
True	False	Total
0.102300[a]	0.897700	1.000000

[a]From Table 6.5, $0.102300 = 1023/10000$

Step 7: Determine Likelihood Probabilities

We select Node $B1$ for this BBN structure as reported in Table 9.202.
We select Node $C2$ for this BBN structure as reported in Table 9.203.

Table 9.202 Node $B1$: likelihood probabilities

$P(B1_j|A_i)$			
$B1_j$			
A_i	True	False	Total
True	0.289345[a]	0.710655	1.000000
False	0.304445	0.695555	1.000000

[a]From Table 6.8, $0.289345 = 296/1023$

Table 9.203 Node $C2$: likelihood probabilities

$P(C2_j|B1_i, A_i)$				
$C2_j$				
$B1_i$	A_i	True	False	Total
True	True	0.481677[a]	0.518323	1.000000
True	False	0.481677	0.518323	1.000000
False	True	0.508679	0.491321	1.000000
False	False	0.508679	0.491321	1.000000

[a]From Table 6.15, $0.481677 = 1459/3029$

We select Node $D2$ for this BBN structure as reported in Table 9.204.

Table 9.204 Node $D2$: likelihood probabilities

$P(D2_j|C2_i, B1_i, A_i)$					
$D2_j$					
$C2_i$	$B1_i$	A_i	True	False	Total
True	True	True	0.692940[a]	0.307060	1.000000
True	True	False	0.692940	0.307060	1.000000
True	False	True	0.717710	0.282290	1.000000
True	False	False	0.717710	0.282290	1.000000
False	True	True	0.715924	0.284076	1.000000
False	True	False	0.715924	0.284076	1.000000
False	False	True	0.716496	0.283504	1.000000
False	False	False	0.716496	0.283504	1.000000

[a]From Table 6.24, $0.692940 = 1011/1459$

Step 8: Determine Joint and Marginal Probabilities

We compute the *joints* and *marginals* for $P(D2_j, C2_i, B1_i, A_i)$ and report these in Table 9.205.

Step 9: Compute Posterior Probabilities

We compute the *posteriors* for $P(C2_j|D2_i, B1_i, A_i)$ and report this in Table 9.206.
We compute the *posteriors* for $P(B1_j|D2_i, C2_i, A_i)$ and report this in Table 9.207.

Table 9.205 Joint and marginal probabilities

$P(D2_j, C2_i, B1_i, A_i)$

			$D2_j$		
$C2_i$	$B1_i$	A_i	True	False	Marginals
True	True	True	0.009880[a]	0.004378	0.014258
True	True	False	0.091220	0.040422	0.131642
True	False	True	0.026542	0.010439	0.036981
True	False	False	0.227958	0.089661	0.317619
False	True	True	0.010984	0.004358	0.015342
False	True	False	0.101416	0.040242	0.141658
False	False	True	0.025593	0.010126	0.035719
False	False	False	0.219807	0.086974	0.306781
Marginals			0.713400	0.286600	1.000000

Note: ZSMS = 0.000000 + 0.000000 = 0.000000
[a]From Tables 9.201, 9.202, 9.203, and 9.204, 0.009880 = 0.692940 * 0.481677 * 0.289345 * 0.102300

Table 9.206 Posterior probabilities

$P(C2_j | D2_i, B1_i, A_i)$

			$C2_j$		
$D2_i$	$B1_i$	A_i	True	False	Total
True	True	True	0.473536[a]	0.526464	1.000000
True	True	False	0.473536	0.526464	1.000000
True	False	True	0.509102	0.490898	1.000000
True	False	False	0.509102	0.490898	1.000000
False	True	True	0.501119	0.498881	1.000000
False	True	False	0.501119	0.498881	1.000000
False	False	True	0.507606	0.492394	1.000000
False	False	False	0.507606	0.492394	1.000000

[a]From Table 9.205, 0.473536 = 0.009880/(0.009880 + 0.010984)

Table 9.207 Posterior probabilities

$P(B1_j | D2_i, C2_i, A_i)$

			$B1_j$		
$D2_i$	$C2_i$	A_i	True	False	Total
True	True	True	0.271262[a]	0.728738	1.000000
True	True	False	0.285797	0.714203	1.000000
True	False	True	0.300301	0.699699	1.000000
True	False	False	0.315718	0.684282	1.000000
False	True	True	0.295462	0.704538	1.000000
False	True	False	0.310741	0.689259	1.000000
False	False	True	0.300893	0.699107	1.000000
False	False	False	0.316327	0.683673	1.000000

[a]From Table 9.205, 0.271262 = 0.009880/(0.009880 + 0.026542)

We compute the *posteriors* for $P(A_j|D2_i, C2_i, B1_i)$ and report this in Table 9.208.

Table 9.208 Posterior probabilities

$P(A_j|D2_i, C2_i, B1_i)$					
A_j					
$D2_i$	$C2_i$	$B1_i$	True	False	Total
True	True	True	0.097722[a]	0.902278	1.000000
True	True	False	0.104289	0.895711	1.000000
True	False	True	0.097722	0.902278	1.000000
True	False	False	0.104289	0.895711	1.000000
False	True	True	0.097722	0.902278	1.000000
False	True	False	0.104289	0.895711	1.000000
False	False	True	0.097722	0.902278	1.000000
False	False	False	0.104289	0.895711	1.000000

[a]From Table 9.205, 0.097722 = 0.009880/(0.009880 + 0.091220)

9.5.11 Path 11: BBN Solution Protocol

Step 2: Specify the BBN–[$A \rightarrow C1|A \rightarrow D4|B2 \rightarrow C1|B2 \rightarrow D4$]

See Fig. 9.27.

Fig. 9.27 Paths
$[A \rightarrow C1|A \rightarrow D4|B2 \rightarrow C1|B2 \rightarrow D4]$

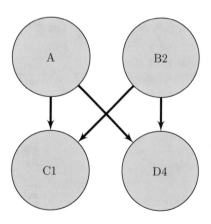

Step 6: Determine Prior or Unconditional Probabilities

We select Node A for this BBN structure as reported in Table 9.209.

Table 9.209 Node A: prior probabilities

$P(A_j)$		
True	False	Total
0.102300[a]	0.897700	1.000000

[a]From Table 6.5, 0.102300 = 1023/10000

Step 7: Determine Likelihood Probabilities

We select Node $B2$ for this BBN structure as reported in Table 9.210.

Table 9.210 Node $B2$: likelihood probabilities

$P(B2_j\|A_i)$			
$B2_j$			
A_i	True	False	Total
True	0.302900[a]	0.697100	1.000000
False	0.302900	0.697100	1.000000

[a]From Table 6.10, $0.302900 = 3029/10000$

We select Node $C1$ for this BBN structure as reported in Table 9.211.

Table 9.211 Node $C1$: likelihood probabilities

$P(C1_j\|B2_i, A_i)$				
$C1_j$				
$B2_i$	A_i	True	False	Total
True	True	0.466216[a]	0.533784	1.000000
True	False	0.483352	0.516648	1.000000
False	True	0.478680	0.521320	1.000000
False	False	0.512172	0.487828	1.000000

[a]From Table 6.13, $0.466216 = 138/296$

We select Node $D4$ for this BBN structure as reported in Table 9.212.

Table 9.212 Node $D4$: likelihood probabilities

$P(D4_j\|C1_i, B2_i, A_i)$					
$D4_j$					
$C1_i$	$B2_i$	A_i	True	False	Total
True	True	True	0.685811[a]	0.314189	1.000000
True	True	False	0.706915	0.293085	1.000000
True	False	True	0.748281	0.251719	1.000000
True	False	False	0.713485	0.286515	1.000000
False	True	True	0.685811	0.314189	1.000000
False	True	False	0.706915	0.293085	1.000000
False	False	True	0.748281	0.251719	1.000000
False	False	False	0.713485	0.286515	1.000000

[a]From Table 6.28, $0.685811 = 203/296$

Step 8: Determine Joint and Marginal Probabilities

We compute the *joints* and *marginals* for $P(D4_j, C1_i, B2_i, A_i)$ and report these in Table 9.213.

Table 9.213 Joint and marginal probabilities

$P(D4_j, C1_i, B2_i, A_i)$					
$D4_j$					
$C1_i$	$B2_i$	A_i	True	False	Marginals
True	True	True	0.009908[a]	0.004539	0.014446
True	True	False	0.092910	0.038520	0.131430
True	False	True	0.025543	0.008593	0.034136
True	False	False	0.228679	0.091831	0.320510
False	True	True	0.011343	0.005197	0.016540
False	True	False	0.099310	0.041174	0.140484
False	False	True	0.027819	0.009358	0.037177
False	False	False	0.217810	0.087466	0.305276
Marginals			0.713322	0.286678	1.000000

Note: ZSMS = 0.000078 − 0.000078 = 0.000000
[a]From Tables 9.209, 9.210, 9.211, and 9.212, 0.009908 = 0.685811 * 0.466216 * 0.302900 * 0.102300

Step 9: Compute Posterior Probabilities

We compute the *posteriors* for $P(C1_j|D4_i, B2_i, A_i)$ and report this in Table 9.214.

Table 9.214 Posterior probabilities

$P(C1_j|D4_i, B2_i, A_i)$					
$C1_j$					
$D4_i$	$B2_i$	A_i	True	False	Total
True	True	True	0.466216[a]	0.533784	1.000000
True	True	False	0.483352	0.516648	1.000000
True	False	True	0.478680	0.521320	1.000000
True	False	False	0.512172	0.487828	1.000000
False	True	True	0.466216	0.533784	1.000000
False	True	False	0.483352	0.516648	1.000000
False	False	True	0.478680	0.521320	1.000000
False	False	False	0.512172	0.487828	1.000000

[a]From Table 9.213, 0.466216 = 0.009908/(0.009908 + 0.011343)

We compute the *posteriors* for $P(B2_j|D4_i, C1_i, A_i)$ and report this in Table 9.215.
We compute the *posteriors* for $P(A_j|D4_i, C1_i, B2_i)$ and report this in Table 9.216.

9.5.12   Path 12: BBN Solution Protocol

Step 2: Specify the BBN–[$A \rightarrow C1|A \rightarrow D3|B2 \rightarrow C1|C1 \rightarrow D3$]

See Fig. 9.28.

Table 9.215 Posterior probabilities

$P(B2_j | D4_i, C1_i, A_i)$

			$B2_j$		
$D4_i$	$C1_i$	A_i	True	False	Total
True	True	True	0.279472[a]	0.720528	1.000000
True	True	False	0.288908	0.711092	1.000000
True	False	True	0.289652	0.710348	1.000000
True	False	False	0.313162	0.686838	1.000000
False	True	True	0.345647	0.654353	1.000000
False	True	False	0.295510	0.704490	1.000000
False	False	True	0.357043	0.642957	1.000000
False	False	False	0.320068	0.679932	1.000000

[a]From Table 9.213, $0.279472 = 0.009908/(0.009908 + 0.025543)$

Table 9.216 Posterior probabilities

$P(A_j | D4_i, C1_i, B2_i)$

			A_j		
$D4_i$	$C1_i$	$B2_i$	True	False	Total
True	True	True	0.096361[a]	0.903639	1.000000
True	True	False	0.100477	0.899523	1.000000
True	False	True	0.102513	0.897487	1.000000
True	False	False	0.113256	0.886744	1.000000
False	True	True	0.105412	0.894588	1.000000
False	True	False	0.085565	0.914435	1.000000
False	False	True	0.112071	0.887929	1.000000
False	False	False	0.096651	0.903349	1.000000

[a]From Table 9.213, $0.096361 = 0.009908/(0.009908 + 0.092910)$

Step 6: Determine Prior or Unconditional Probabilities

We select Node *A* for this BBN structure as reported in Table 9.217.

Table 9.217 Node *A*: prior probabilities

$P(A_j)$

True	False	Total
0.102300[a]	0.897700	1.000000

[a]From Table 6.5, $0.102300 = 1023/10000$

Step 7: Determine Likelihood Probabilities

We select Node *B*2 for this BBN structure as reported in Table 9.218.

Fig. 9.28 Paths
$[A \rightarrow C1|A \rightarrow D3|B2 \rightarrow C1|C1 \rightarrow D3]$

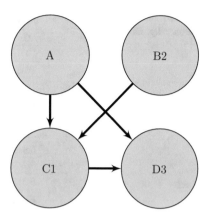

Table 9.218 Node $B2$:
likelihood probabilities

$P(B2_j|A_i)$			
$B2_j$			
A_i	True	False	Total
True	0.302900[a]	0.697100	1.000000
False	0.302900	0.697100	1.000000

[a]From Table 6.10, $0.302900 = 3029/10000$

We select Node $C1$ for this BBN structure as reported in Table 9.219.

Table 9.219 Node $C1$:
likelihood probabilities

$P(C1_j|B2_i, A_i)$				
$C1_j$				
$B2_i$	A_i	True	False	Total
True	True	0.466216[a]	0.533784	1.000000
True	False	0.483352	0.516648	1.000000
False	True	0.478680	0.521320	1.000000
False	False	0.512172	0.487828	1.000000

[a]From Table 6.13, $0.466216 = 138/296$

We select Node $D3$ for this BBN structure as reported in Table 9.220.

Step 8: Determine Joint and Marginal Probabilities

We compute the *joints* and *marginals* for $P(D3_j, C1_i, B2_i, A_i)$ and report these in Table 9.221.

Table 9.220 Node $D3$: likelihood probabilities

$P(D3_j|C1_i, B2_i, A_i)$					
$D3_j$					
$C1_i$	$B2_i$	A_i	True	False	Total
True	True	True	0.716049[a]	0.283951	1.000000
True	True	False	0.709892	0.290108	1.000000
True	False	True	0.716049	0.283951	1.000000
True	False	False	0.709892	0.290108	1.000000
False	True	True	0.743017	0.256983	1.000000
False	True	False	0.713100	0.286900	1.000000
False	False	True	0.743017	0.256983	1.000000
False	False	False	0.713100	0.286900	1.000000

[a]From Table 6.26, $0.716049 = 348/486$

Table 9.221 Joint and marginal probabilities

$P(D3_j, C1_i, B2_i, A_i)$					
$D3_j$					
$C1_i$	$B2_i$	A_i	True	False	Marginals
True	True	True	0.010344[a]	0.004102	0.014446
True	True	False	0.093301	0.038129	0.131430
True	False	True	0.024443	0.009693	0.034136
True	False	False	0.227527	0.092983	0.320510
False	True	True	0.012290	0.004251	0.016540
False	True	False	0.100179	0.040305	0.140484
False	False	True	0.027623	0.009554	0.037177
False	False	False	0.217693	0.087584	0.305276
Marginals			0.713400	0.286600	1.000000

Note: ZSMS = 0.000000 + 0.000000 = 0.000000
[a]From Tables 9.217, 9.218, 9.219, and 9.220, 0.010344 = 0.716049 * 0.466216 * 0.302900 * 0.102300

Step 9: Compute Posterior Probabilities

We compute the *posteriors* for $P(C1_j|D3_i, B2_i, A_i)$ and report this in Table 9.222.
We compute the *posteriors* for $P(B2_j|D3_i, C1_i, A_i)$ and report this in Table 9.223.
We compute the *posteriors* for $P(A_j|D3_i, C1_i, B2_i)$ and report this in Table 9.224.

9.5.13   Path 13: BBN Solution Protocol

Step 2: Specify the BBN–[$A \to C3|A \to D1|B2 \to D1|C3 \to D1$]

See Fig. 9.29.

Table 9.222 Posterior probabilities

$P(C1_j | D3_i, B2_i, A_i)$

			$C1_j$		
$D3_i$	$B2_i$	A_i	True	False	Total
True	True	True	0.457029[a]	0.542971	1.000000
True	True	False	0.482226	0.517774	1.000000
True	False	True	0.469462	0.530538	1.000000
True	False	False	0.511045	0.488955	1.000000
False	True	True	0.491113	0.508887	1.000000
False	True	False	0.486129	0.513871	1.000000
False	False	True	0.503614	0.496386	1.000000
False	False	False	0.514950	0.485050	1.000000

[a]From Table 9.221, $0.457029 = 0.010344/(0.010344 + 0.012290)$

Table 9.223 Posterior probabilities

$P(B2_j | D3_i, C1_i, A_i)$

			$B2_j$		
$D3_i$	$C1_i$	A_i	True	False	Total
True	True	True	0.297359[a]	0.702641	1.000000
True	True	False	0.290812	0.709188	1.000000
True	False	True	0.307912	0.692088	1.000000
True	False	False	0.315155	0.684845	1.000000
False	True	True	0.297359	0.702641	1.000000
False	True	False	0.290812	0.709188	1.000000
False	False	True	0.307912	0.692088	1.000000
False	False	False	0.315155	0.684845	1.000000

[a]From Table 9.221, $0.297359 = 0.010344/(0.010344 + 0.024443)$

Table 9.224 Posterior probabilities

$P(A_j | D3_i, C1_i, B2_i)$

			A_j		
$D3_i$	$C1_i$	$B2_i$	True	False	Total
True	True	True	0.099806	0.900194	1.000000
True	True	False	0.097008	0.902992	1.000000
True	False	True	0.109272	0.890728	1.000000
True	False	False	0.112603	0.887397	1.000000
False	True	True	0.097135	0.902865	1.000000
False	True	False	0.094404	0.905596	1.000000
False	False	True	0.095399	0.904601	1.000000
False	False	False	0.098354	0.901646	1.000000

[a]From Table 9.221, $0.099806 = 0.010344/(0.010344 + 0.093301)$

Step 6: Determine Prior or Unconditional Probabilities

We select Node A for this BBN structure as reported in Table 9.225.

Fig. 9.29 Paths
$[A \rightarrow C3|A \rightarrow D1|B2 \rightarrow D1|C3 \rightarrow D1]$

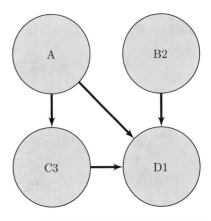

Table 9.225 Node A: prior
probabilities

P(A)		
True	False	Total
0.102300[a]	0.897700	1.000000

[a]From Table 6.5, 0.102300 = 1023/10000

Step 7: Determine Likelihood Probabilities

We select Node $B2$ for this BBN structure as reported in Table 9.226.

Table 9.226 Node $B2$:
likelihood probabilities

$P(B2_j|A_i)$			
$B2_j$			
A_i	True	False	Total
True	0.302900[a]	0.697100	1.000000
False	0.302900	0.697100	1.000000

[a]From Table 6.10, 0.302900 = 3029/10000

We select Node $C3$ for this BBN structure as reported in Table 9.227.

Table 9.227 Node $C3$:
likelihood probabilities

$P(C3_j|B2_i, A_i)$				
$C3_j$				
$B2_i$	A_i	True	False	Total
True	True	0.475073[a]	0.524927	1.000000
True	False	0.503398	0.496602	1.000000
False	True	0.475073	0.524927	1.000000
False	False	0.503398	0.496602	1.000000

[a]From Table 6.17, 0.475073 = 486/1023

We select Node $D1$ for this BBN structure as reported in Table 9.228.

Table 9.228 Node $D1$: likelihood probabilities

$P(D1_j | C3_i, B2_i, A_i)$

$D1_j$					
$C3_i$	$B2_i$	A_i	True	False	Total
True	True	True	0.637681[a]	0.362319	1.000000
True	True	False	0.698713	0.301287	1.000000
True	False	True	0.747126	0.252874	1.000000
True	False	False	0.714509	0.285491	1.000000
False	True	True	0.727848	0.272152	1.000000
False	True	False	0.714589	0.285411	1.000000
False	False	True	0.749340	0.250660	1.000000
False	False	False	0.712410	0.287590	1.000000

[a]From Table 6.22, $0.637681 = 88/138$

Step 8: Determine Joint and Marginal Probabilities

We compute the *joints* and *marginals* for $P(D1_j, C3_i, B2_i, A_i)$ and report these in Table 9.229.

Table 9.229 Joint and marginal probabilities

$P(D1_j, C3_i, B2_i, A_i)$

$D1_j$					
$C3_i$	$B2_i$	A_i	True	False	Marginals
True	True	True	0.009387[a]	0.005334	0.014721
True	True	False	0.095640	0.041240	0.136881
True	False	True	0.025312	0.008567	0.033879
True	False	False	0.225084	0.089935	0.315019
False	True	True	0.011839	0.004427	0.016266
False	True	False	0.096493	0.038540	0.135033
False	False	True	0.028051	0.009383	0.037434
False	False	False	0.221394	0.089374	0.310767
Marginals			0.713200	0.286800	1.000000

Note: ZSMS $= 0.000200 - 0.000200 = 0.000000$
[a]From Tables 9.225, 9.226, 9.227, and 9.228, $0.009387 = 0.637681 * 0.475073 * 0.302900 * 0.102300$

Step 9: Compute Posterior Probabilities

We compute the *posteriors* for $P(C3_j | D1_i, B2_i, A_i)$ and report this in Table 9.230.
We compute the *posteriors* for $P(B2_j | D1_i, C3_i, A_i)$ and report this in Table 9.231.
We compute the *posteriors* for $P(A_j | D1_i, C3_i, B2_i)$ and report this in Table 9.232.

Table 9.230 Posterior
probabilities

$P(C3_j | D1_i, B2_i, A_i)$

| | | | $C3_j$ | | |
$D1_i$	$B2_i$	A_i	True	False	Total
True	True	True	0.442248[a]	0.557752	1.000000
True	True	False	0.497781	0.502219	1.000000
True	False	True	0.474335	0.525665	1.000000
True	False	False	0.504133	0.495867	1.000000
False	True	True	0.546459	0.453541	1.000000
False	True	False	0.516925	0.483075	1.000000
False	False	True	0.477267	0.522733	1.000000
False	False	False	0.501566	0.498434	1.000000

[a]From Table 9.229, 0.442248 = 0.009387/(0.009387 + 0.011839)

Table 9.231 Posterior
probabilities

$P(B2_j | D1_i, C3_i, A_i)$

| | | | $B2_j$ | | |
$D1_i$	$C3_i$	A_i	True	False	Total
True	True	True	0.270533[a]	0.729467	1.000000
True	True	False	0.298201	0.701799	1.000000
True	False	True	0.296791	0.703209	1.000000
True	False	False	0.303545	0.696455	1.000000
False	True	True	0.383696	0.616304	1.000000
False	True	False	0.314390	0.685610	1.000000
False	False	True	0.320546	0.679454	1.000000
False	False	False	0.301296	0.698704	1.000000

[a]From Table 9.229, 0.270533 = 0.009387/(0.009387 + 0.025312)

Table 9.232 Posterior
probabilities

$P(A_j | D1_i, C3_i, B2_i)$

| | | | A_j | | |
$D1_i$	$C3_i$	$B2_i$	True	False	Total
True	True	True	0.089379[a]	0.910621	1.000000
True	True	False	0.101088	0.898912	1.000000
True	False	True	0.109284	0.890716	1.000000
True	False	False	0.112454	0.887546	1.000000
False	True	True	0.114520	0.885480	1.000000
False	True	False	0.086974	0.913026	1.000000
False	False	True	0.103028	0.896972	1.000000
False	False	False	0.095014	0.904986	1.000000

[a]From Table 9.229, 0.089379 = 0.009387/(0.009387 + 0.095640)

Fig. 9.30 Paths
$[A \rightarrow C1|B2 \rightarrow C1|B2 \rightarrow D2|C1 \rightarrow D2]$

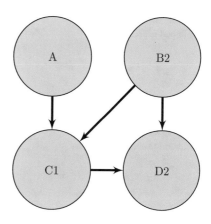

9.5.14   Path 14: BBN Solution Protocol

Step 2: Specify the BBN–[$A \rightarrow C1|B2 \rightarrow C1|B2 \rightarrow D2|C1 \rightarrow D2$]

See Fig. 9.30.

Step 6: Determine Prior or Unconditional Probabilities

We select Node A for this BBN structure as reported in Table 9.233.

Table 9.233 Node A: prior probabilities

P(A)		
True	False	Total
0.102300[a]	0.897700	1.000000

[a]From Table 6.5, 0.102300 = 1023/10000

Step 7: Determine Likelihood Probabilities

We select Node $B2$ for this BBN structure as reported in Table 9.234.
We select Node $C1$ for this BBN structure as reported in Table 9.235.
We select Node $D2$ for this BBN structure as reported in Table 9.236.

Step 8: Determine Joint and Marginal Probabilities

We compute the *joints* and *marginals* for $P(D2_j, C1_i, B2_i, A_i)$ and report these in Table 9.237.

Table 9.234 Node $B2$: likelihood probabilities

$P(B2_j\|A_i)$			
$B2_j$			
A_i	True	False	Total
True	0.302900[a]	0.697100	1.000000
False	0.302900	0.697100	1.000000

[a]From Table 6.10, 0.302900 = 3029/10000

Table 9.235 Node $C1$: likelihood probabilities

$P(C1_j\|B2_i, A_i)$				
$C1_j$				
$B2_i$	A_i	True	False	Total
True	True	0.466216[a]	0.533784	1.000000
True	False	0.483352	0.516648	1.000000
False	True	0.478680	0.521320	1.000000
False	False	0.512172	0.487828	1.000000

[a]From Table 6.13, 0.466216 = 138/296

Table 9.236 Node $D2$: likelihood probabilities

$P(D2_j\|C1_i, B2_i, A_i)$					
$D2_j$					
$C1_i$	$B2_i$	A_i	True	False	Total
True	True	True	0.692940[a]	0.307060	1.000000
True	True	False	0.692940	0.307060	1.000000
True	False	True	0.717710	0.282290	1.000000
True	False	False	0.717710	0.282290	1.000000
False	True	True	0.715924	0.284076	1.000000
False	True	False	0.715924	0.284076	1.000000
False	False	True	0.716496	0.283504	1.000000
False	False	False	0.716496	0.283504	1.000000

[a]From Table 6.24, 0.692940 = 1011/1459

Step 9: Compute Posterior Probabilities

We compute the *posteriors* for $P(C1_j|D2_i, B2_i, A_i)$ and report this in Table 9.238.
We compute the *posteriors* for $P(B2_j|D2_i, C1_i, A_i)$ and report this in Table 9.239.
We compute the *posteriors* for $P(A_j|D2_i, C1_i, B2_i)$ and report this in Table 9.240.

9.5.15   Path 15: BBN Solution Protocol

Step 2: Specify the BBN–[$A \rightarrow D1|B2 \rightarrow C2|B2 \rightarrow D1|C2 \rightarrow D1$]

See Fig. 9.31.

Table 9.237 Joint and marginal probabilities

$P(D2_j, C1_i, B2_i, A_i)$

$D2_j$

$C1_i$	$B2_i$	A_i	True	False	Marginals
True	True	True	0.010011[a]	0.004436	0.014446
True	True	False	0.091073	0.040357	0.131430
True	False	True	0.024500	0.009636	0.034136
True	False	False	0.230033	0.090477	0.320510
False	True	True	0.011842	0.004699	0.016540
False	True	False	0.100576	0.039908	0.140484
False	False	True	0.026637	0.010540	0.037177
False	False	False	0.218729	0.086547	0.305276
Marginals			0.713401	0.286599	1.000000

Note: ZSMS $= -0.000001 + 0.000001 = 0.000000$
[a]From Tables 9.233, 9.234, 9.235, and 9.236, $0.010011 = 0.692940 * 0.466216 * 0.302900 * 0.102300$

Table 9.238 Posterior probabilities

$P(C1_j | D2_i, B2_i, A_i)$

$C1_j$

$D2_i$	$B2_i$	A_i	True	False	Total
True	True	True	0.458106[a]	0.541894	1.000000
True	True	False	0.475208	0.524792	1.000000
True	False	True	0.479102	0.520898	1.000000
True	False	False	0.512595	0.487405	1.000000
False	True	True	0.485618	0.514382	1.000000
False	True	False	0.502795	0.497205	1.000000
False	False	True	0.477609	0.522391	1.000000
False	False	False	0.511100	0.488900	1.000000

[a]From Table 9.237, $0.458106 = 0.010011/(0.010011 + 0.011842)$

Table 9.239 Posterior probabilities

$P(B2_j | D2_i, C1_i, A_i)$

$B2_j$

$D2_i$	$C1_i$	A_i	True	False	Total
True	True	True	0.290073[a]	0.709927	1.000000
True	True	False	0.283622	0.716378	1.000000
True	False	True	0.307741	0.692259	1.000000
True	False	False	0.314983	0.685017	1.000000
False	True	True	0.315226	0.684774	1.000000
False	True	False	0.308459	0.691541	1.000000
False	False	True	0.308342	0.691658	1.000000
False	False	False	0.315591	0.684409	1.000000

[a]From Table 9.237, $0.290073 = 0.010011/(0.010011 + 0.024500)$

Table 9.240 Posterior probabilities

			$P(A_j|D2_i, C1_i, B2_i)$		
A_j					
$D2_i$	$C1_i$	$B2_i$	True	False	Total
True	True	True	0.099032[a]	0.900968	1.000000
True	True	False	0.096254	0.903746	1.000000
True	False	True	0.105336	0.894664	1.000000
True	False	False	0.108561	0.891439	1.000000
False	True	True	0.099032	0.900968	1.000000
False	True	False	0.096254	0.903746	1.000000
False	False	True	0.105336	0.894664	1.000000
False	False	False	0.108561	0.891439	1.000000

[a]From Table 9.237, $0.099032 = 0.010011/(0.010011 + 0.091073)$

Fig. 9.31 Paths
$[A \rightarrow D1|B2 \rightarrow C2|B2 \rightarrow D1|C2 \rightarrow D1]$

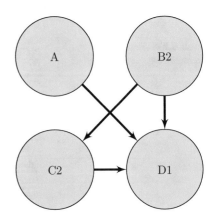

Step 6: Determine Prior or Unconditional Probabilities

We select Node A for this BBN structure as reported in Table 9.241.

Table 9.241 Node A: prior probabilities

$P(A_j)$		
True	False	Total
0.102300[a]	0.897700	1.000000

[a]From Table 6.5, $0.102300 = 1023/10000$

Step 7: Determine Likelihood Probabilities

We select Node $B2$ for this BBN structure as reported in Table 9.242.
We select Node $C2$ for this BBN structure as reported in Table 9.243.

Table 9.242 Node $B2$: likelihood probabilities

$P(B2_j \mid A_i)$			
$B2_j$			
A_i	True	False	Total
~~True~~	0.302900[a]	0.697100	1.000000
~~False~~	~~0.302900~~	~~0.697100~~	~~1.000000~~

[a]From Table 6.10, 0.302900 = 3029/10000

Table 9.243 Node $C2$: likelihood probabilities

$P(C2_j \mid B2_i, A_i)$				
$C2_j$				
$B2_i$	A_i	True	False	Total
True	~~True~~	0.481677[a]	0.518323	1.000000
~~True~~	~~False~~	~~0.481677~~	~~0.518323~~	~~1.000000~~
False	~~True~~	0.508679	0.491321	1.000000
~~False~~	~~False~~	~~0.508679~~	~~0.491321~~	~~1.000000~~

[a]From Table 6.15, 0.481677 = 1459/3029

We select Node $D1$ for this BBN structure as reported in Table 9.244.

Table 9.244 Node $D1$: likelihood probabilities

$P(D1_j \mid C2_i, B2_i, A_i)$					
$D1_j$					
$C2_i$	$B2_i$	A_i	True	False	Total
True	True	True	0.637681[a]	0.362319	1.000000
True	True	False	0.698713	0.301287	1.000000
True	False	True	0.747126	0.252874	1.000000
True	False	False	0.714509	0.285491	1.000000
False	True	True	0.727848	0.272152	1.000000
False	True	False	0.714589	0.285411	1.000000
False	False	True	0.749340	0.250660	1.000000
False	False	False	0.712410	0.287590	1.000000

[a]From Table 6.22, 0.637681 = 88/138

Step 8: Determine Joint and Marginal Probabilities

We compute the *joints* and *marginals* for $P(D1_j, C2_i, B2_i, A_i)$ and report these in Table 9.245.

Step 9: Compute Posterior Probabilities

We compute the *posteriors* for $P(C2_j \mid D1_i, B2_i, A_i)$ and report this in Table 9.246.

Table 9.245 Joint and
marginal probabilities

$P(D1_j, C2_i, B2_i, A_i)$

$D1_j$

$C2_i$	$B2_i$	A_i	True	False	Marginals
True	True	True	0.009518[a]	0.005408	0.014926
True	True	False	0.091514	0.039461	0.130974
True	False	True	0.027102	0.009173	0.036276
True	False	False	0.227446	0.090879	0.318324
False	True	True	0.011690	0.004371	0.016061
False	True	False	0.100713	0.040225	0.140939
False	False	True	0.026255	0.008783	0.035038
False	False	False	0.219039	0.088423	0.307462
Marginals			0.713277	0.286723	1.000000

Note: ZSMS $= 0.000123 - 0.000123 = 0.000000$
[a]From Tables 9.241, 9.242, 9.243, and 9.244, $0.009518 = 0.637681 * 0.481677 * 0.302900 * 0.102300$

Table 9.246 Posterior
probabilities

$P(C2_j|D1_i, B2_i, A_i)$

$C2_j$

$D1_i$	$B2_i$	A_i	True	False	Total
True	True	True	0.448786[a]	0.551214	1.000000
True	True	False	0.476070	0.523930	1.000000
True	False	True	0.507939	0.492061	1.000000
True	False	False	0.509414	0.490586	1.000000
False	True	True	0.553010	0.446990	1.000000
False	True	False	0.495202	0.504798	1.000000
False	False	True	0.510876	0.489124	1.000000
False	False	False	0.506848	0.493152	1.000000

[a]From Table 9.245, $0.448786 = 0.009518/(0.009518 + 0.011690)$

We compute the *posteriors* for $P(B2_j|D1_i, C2_i, A_i)$ and report this in Table 9.247.
We compute the *posteriors* for $P(A_j|D1_i, C2_i, B2_i)$ and report this in Table 9.248.

9.6   5-Path BBN

9.6.1   Path 1: BBN Solution Protocol

Step 2: Specify the BBN–[$A \rightarrow B1|A \rightarrow C1|A \rightarrow D4|B1 \rightarrow C1|B1 \rightarrow D4$]

See Fig. 9.32.

Table 9.247 Posterior probabilities

$P(B2_j\|D1_i, C2_i, A_i)$					
			$B2_j$		
$D1_i$	$C2_i$	A_i	True	False	Total
True	True	True	0.259905[a]	0.740095	1.000000
True	True	False	0.286913	0.713087	1.000000
True	False	True	0.308077	0.691923	1.000000
True	False	False	0.314973	0.685027	1.000000
False	True	True	0.370882	0.629118	1.000000
False	True	False	0.302754	0.697246	1.000000
False	False	True	0.332309	0.667691	1.000000
False	False	False	0.312677	0.687323	1.000000

[a]From Table 9.245, $0.259905 = 0.009518/(0.009518 + 0.027102)$

Table 9.248 Posterior probabilities

$P(A_j\|D1_i, C2_i, B2_i)$					
			A_j		
$D1_i$	$C2_i$	$B2_i$	True	False	Total
True	True	True	0.094206[a]	0.905794	1.000000
True	True	False	0.106473	0.893527	1.000000
True	False	True	0.104001	0.895999	1.000000
True	False	False	0.107036	0.892964	1.000000
False	True	True	0.120525	0.879475	1.000000
False	True	False	0.091684	0.908316	1.000000
False	False	True	0.098013	0.901987	1.000000
False	False	False	0.090350	0.909650	1.000000

[a]From Table 9.245, $0.094206 = 0.009518/(0.009518 + 0.091514)$

Step 6: Determine Prior or Unconditional Probabilities

We select Node A for this BBN structure as reported in Table 9.249.

Table 9.249 Node A: prior probabilities

$P(A_j)$		
True	False	Total
0.102300[a]	0.897700	1.000000

[a]From Table 6.5, $0.102300 = 1023/10000$

Step 7: Determine Likelihood Probabilities

We select Node $B1$ for this BBN structure as reported in Table 9.250.

Fig. 9.32 Paths
$[A \to B1|A \to C1|A \to D4|B1 \to C1|B1 \to D4]$

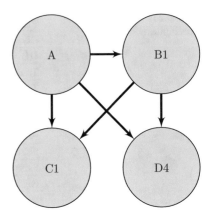

Table 9.250 Node $B1$:
likelihood probabilities

$P(B1_j|A_i)$			
$B1_j$			
A_i	True	False	Total
True	0.289345[a]	0.710655	1.000000
False	0.304445	0.695555	1.000000

[a]From Table 6.8, $0.289345 = 296/1023$

We select Node $C1$ for this BBN structure as reported in Table 9.251.

Table 9.251 Node $C1$:
likelihood probabilities

$P(C1_j|B1_i, A_i)$				
$C1_j$				
$B1_i$	A_i	True	False	Total
True	True	0.466216[a]	0.533784	1.000000
True	False	0.483352	0.516648	1.000000
False	True	0.478680	0.521320	1.000000
False	False	0.512172	0.487828	1.000000

[a]From Table 6.13, $0.466216 = 138/296$

We select Node $D4$ for this BBN structure as reported in Table 9.252.

Step 8: Determine Joint and Marginal Probabilities

We compute the *joints* and *marginals* for $P(D4_j, C1_i, B1_i, A_i)$ and report these in Table 9.253.

Table 9.252 Node $D4$: likelihood probabilities

$P(D4_j | C1_i, B1_i, A_i)$

$D4_j$					
$C1_i$	$B1_i$	A_i	True	False	Total
~~True~~	True	True	0.685811[a]	0.314189	1.000000
~~True~~	True	False	0.706915	0.293085	1.000000
~~True~~	False	True	0.748281	0.251719	1.000000
~~True~~	False	False	0.713485	0.286515	1.000000
~~False~~	~~True~~	~~True~~	~~0.685811~~	~~0.314189~~	~~1.000000~~
~~False~~	~~True~~	~~False~~	~~0.706915~~	~~0.293085~~	~~1.000000~~
~~False~~	False	~~True~~	~~0.748281~~	~~0.251719~~	~~1.000000~~
~~False~~	~~False~~	~~False~~	~~0.713485~~	~~0.286515~~	~~1.000000~~

[a]From Table 6.28, $0.685811 = 203/296$

Table 9.253 Joint and marginal probabilities

$P(D4_j, C1_i, B1_i, A_i)$

$D4_j$					
$C1_i$	$B1_i$	A_i	True	False	Marginals
True	True	True	0.009464[a]	0.004336	0.013800
True	True	False	0.093384	0.038716	0.132100
True	False	True	0.026040	0.008760	0.034800
True	False	False	0.228172	0.091628	0.319800
False	True	True	0.010836	0.004964	0.015800
False	True	False	0.099816	0.041384	0.141200
False	False	True	0.028360	0.009540	0.037900
False	False	False	0.217328	0.087272	0.304600
Marginals			0.713400	0.286600	1.000000

Note: ZSMS $= 0.000000 + 0.000000 = 0.000000$
[a]From Tables 9.249, 9.250, 9.251, and 9.252, $0.009464 = 0.685811 * 0.466216 * 0.289345 * 0.102300$

Step 9: Compute Posterior Probabilities

We compute the *posteriors* for $P(C1_j | D4_i, B1_i, A_i)$ and report this in Table 9.254.
We compute the *posteriors* for $P(B1_j | D4_i, C1_i, A_i)$ and report this in Table 9.255.
We compute the *posteriors* for $P(A_j | D4_i, C1_i, B1_i)$ and report this in Table 9.256.

9.6.2 Path 2: BBN Solution Protocol

Step 2: Specify the BBN–$[A \rightarrow B1 | A \rightarrow C1 | A \rightarrow D3 | B1 \rightarrow C1 | C1 \rightarrow D3]$

See Fig. 9.33.

Table 9.254 Posterior
probabilities

			$P(C1_j\|D4_i, B1_i, A_i)$		
			$C1_j$		
$D4_i$	$B1_i$	A_i	True	False	Total
True	True	True	0.466216[a]	0.533784	1.000000
True	True	False	0.483352	0.516648	1.000000
True	False	True	0.478680	0.521320	1.000000
True	False	False	0.512172	0.487828	1.000000
False	True	True	0.466216	0.533784	1.000000
False	True	False	0.483352	0.516648	1.000000
False	False	True	0.478680	0.521320	1.000000
False	False	False	0.512172	0.487828	1.000000

[a]From Table 9.253, 0.466216 = 0.009464/(0.009464 + 0.010836)

Table 9.255 Posterior
probabilities

			$P(B1_j\|D4_i, C1_i, A_i)$		
			$B1_j$		
$D4_i$	$C1_i$	A_i	True	False	Total
True	True	True	0.266564[a]	0.733436	1.000000
True	True	False	0.290411	0.709589	1.000000
True	False	True	0.276454	0.723546	1.000000
True	False	False	0.314735	0.685265	1.000000
False	True	True	0.331088	0.668912	1.000000
False	True	False	0.297033	0.702967	1.000000
False	False	True	0.342255	0.657745	1.000000
False	False	False	0.321660	0.678340	1.000000

[a]From Table 9.253, 0.266564 = 0.009464/(0.009464 + 0.026040)

Table 9.256 Posterior
probabilities

			$P(A_j\|D4_i, C1_i, B1_i)$		
			A_j		
$D4_i$	$C1_i$	$B1_i$	True	False	Total
True	True	True	0.092021[a]	0.907979	1.000000
True	True	False	0.102435	0.897565	1.000000
True	False	True	0.097927	0.902073	1.000000
True	False	False	0.115431	0.884569	1.000000
False	True	True	0.100710	0.899290	1.000000
False	True	False	0.087260	0.912740	1.000000
False	False	True	0.107108	0.892892	1.000000
False	False	False	0.098543	0.901457	1.000000

[a]From Table 9.253, 0.092021 = 0.009464/(0.009464 + 0.093384)

Step 6: Determine Prior or Unconditional Probabilities

We select Node A for this BBN structure as reported in Table 9.257.

Fig. 9.33 Paths
$[A \rightarrow B1|A \rightarrow C1|A \rightarrow D3|B1 \rightarrow C1|C1 \rightarrow D3]$

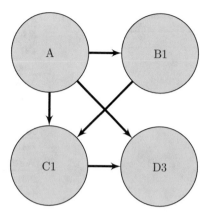

Table 9.257 Node A: prior probabilities

$P(A_j)$		
True	False	Total
0.102300[a]	0.897700	1.000000

[a]From Table 6.5, $0.102300 = 1023/10000$

Step 7: Determine Likelihood Probabilities

We select Node $B1$ for this BBN structure as reported in Table 9.258.

Table 9.258 Node $B1$: likelihood probabilities

$P(B1_j|A_i)$			
$B1_j$			
A_i	True	False	Total
True	0.289345[a]	0.710655	1.000000
False	0.304445	0.695555	1.000000

[a]From Table 6.8, $0.289345 = 296/1023$

We select Node $C1$ for this BBN structure as reported in Table 9.259.

Table 9.259 Node $C1$: likelihood probabilities

$P(C1_j|B1_i, A_i)$				
$C1_j$				
$B1_i$	A_i	True	False	Total
True	True	0.466216[a]	0.533784	1.000000
True	False	0.483352	0.516648	1.000000
False	True	0.478680	0.521320	1.000000
False	False	0.512172	0.487828	1.000000

[a]From Table 6.13, $0.466216 = 138/296$

We select Node $D3$ for this BBN structure as reported in Table 9.260.

Table 9.260 Node $D3$: likelihood probabilities

$P(D3_j | C1_i, B1_i, A_i)$

$C1_i$	$B1_i$	A_i	True	False	Total
True	True	True	0.716049[a]	0.283951	1.000000
True	True	False	0.709892	0.290108	1.000000
True	False	True	0.716049	0.283951	1.000000
True	False	False	0.709892	0.290108	1.000000
False	True	True	0.743017	0.256983	1.000000
False	True	False	0.713100	0.286900	1.000000
False	False	True	0.743017	0.256983	1.000000
False	False	False	0.713100	0.286900	1.000000

[a]From Table 6.26, $0.716049 = 348/486$

Step 8: Determine Joint and Marginal Probabilities

We compute the *joints* and *marginals* for $P(D3_j, C1_i, B1_i, A_i)$ and report these in Table 9.261.

Table 9.261 Joint and marginal probabilities

$P(D3_j, C1_i, B1_i, A_i)$

$C1_i$	$B1_i$	A_i	True	False	Marginals
True	True	True	0.009881[a]	0.003919	0.013800
True	True	False	0.093777	0.038323	0.132100
True	False	True	0.024919	0.009881	0.034800
True	False	False	0.227023	0.092777	0.319800
False	True	True	0.011740	0.004060	0.015800
False	True	False	0.100690	0.040510	0.141200
False	False	True	0.028160	0.009740	0.037900
False	False	False	0.217210	0.087390	0.304600
Marginals			0.713400	0.286600	1.000000

Note: ZSMS $= 0.000000 + 0.000000 = 0.000000$
[a]From Tables 9.257, 9.258, 9.259, and 9.260, $0.009881 = 0.716049 * 0.466216 * 0.289345 * 0.102300$

Step 9: Compute Posterior Probabilities

We compute the *posteriors* for $P(C1_j | D3_i, B1_i, A_i)$ and report this in Table 9.262.
We compute the *posteriors* for $P(B1_j | D3_i, C1_i, A_i)$ and report this in Table 9.263.
We compute the *posteriors* for $P(A_j | D3_i, C1_i, B1_i)$ and report this in Table 9.264.

Table 9.262 Posterior probabilities

$P(C1_j\|D3_i, B1_i, A_i)$					
			$C1_j$		
$D3_i$	$B1_i$	A_i	True	False	Total
True	True	True	0.457029[a]	0.542971	1.000000
True	True	False	0.482226	0.517774	1.000000
True	False	True	0.469462	0.530538	1.000000
True	False	False	0.511045	0.488955	1.000000
False	True	True	0.491113	0.508887	1.000000
False	True	False	0.486129	0.513871	1.000000
False	False	True	0.503614	0.496386	1.000000
False	False	False	0.514950	0.485050	1.000000

[a]From Table 9.261, 0.457029 = 0.009881/(0.009881 + 0.011740)

Table 9.263 Posterior probabilities

$P(B1_j\|D3_i, C1_i, A_i)$					
			$B1_j$		
$D3_i$	$C1_i$	A_i	True	False	Total
True	True	True	0.283951[a]	0.716049	1.000000
True	True	False	0.292321	0.707679	1.000000
True	False	True	0.294227	0.705773	1.000000
True	False	False	0.316734	0.683266	1.000000
False	True	True	0.283951	0.716049	1.000000
False	True	False	0.292321	0.707679	1.000000
False	False	True	0.294227	0.705773	1.000000
False	False	False	0.316734	0.683266	1.000000

[a]From Table 9.261, 0.283951 = 0.009881/(0.009881 + 0.024919)

Table 9.264 Posterior probabilities

$P(A_j\|D3_i, C1_i, B1_i)$					
			A_j		
$D3_i$	$C1_i$	$B1_i$	True	False	Total
True	True	True	0.095328[a]	0.904672	1.000000
True	True	False	0.098906	0.901094	1.000000
True	False	True	0.104418	0.895582	1.000000
True	False	False	0.114767	0.885233	1.000000
False	True	True	0.092764	0.907236	1.000000
False	True	False	0.096256	0.903744	1.000000
False	False	True	0.091099	0.908901	1.000000
False	False	False	0.100275	0.899725	1.000000

[a]From Table 9.261, 0.095328 = 0.009881/(0.009881 + 0.093777)

Fig. 9.34 Paths
$[A \rightarrow B1 | A \rightarrow C3 | A \rightarrow D1 | B1 \rightarrow D1 | C3 \rightarrow D1]$

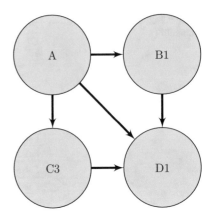

9.6.3 Path 3: BBN Solution Protocol

Step 2: Specify the BBN–$[A \rightarrow B1 | A \rightarrow C3 | A \rightarrow D1 | B1 \rightarrow D1 | C3 \rightarrow D1]$

See Fig. 9.34.

Step 6: Determine Prior or Unconditional Probabilities

We select Node A for this BBN structure as reported in Table 9.265.

Table 9.265 Node A: prior probabilities

$P(A_j)$		
True	False	Total
0.102300[a]	0.897700	1.000000

[a]From Table 6.5, $0.102300 = 1023/10000$

Step 7: Determine Likelihood Probabilities

We select Node $B1$ for this BBN structure as reported in Table 9.266.
We select Node $C3$ for this BBN structure as reported in Table 9.267.
We select Node $D1$ for this BBN structure as reported in Table 9.268.

Step 8: Determine Joint and Marginal Probabilities

We compute the *joints* and *marginals* for $P(D1_j, C3_i, B1_i, A_i)$ and report these in Table 9.269.

Table 9.266 Node $B1$:
likelihood probabilities

	$P(B1_j|A_i)$		
	$B1_j$		
A_i	True	False	Total
True	0.289345[a]	0.710655	1.000000
False	0.304445	0.695555	1.000000

[a]From Table 6.8, 0.289345 = 296/1023

Table 9.267 Node $C3$:
likelihood probabilities

		$P(C3_j|\cancel{B1}_i, A_i)$		
		$C3_j$		
$\cancel{B1}_i$	A_i	True	False	Total
~~True~~	True	0.475073[a]	0.524927	1.000000
~~True~~	False	0.503398	0.496602	1.000000
~~False~~	~~True~~	~~0.475073~~	~~0.524927~~	~~1.000000~~
~~False~~	~~False~~	~~0.503398~~	~~0.496602~~	~~1.000000~~

[a]From Table 6.17, 0.475073 = 486/1023

Table 9.268 Node $D1$:
likelihood probabilities

			$P(D1_j|C3_i, B1_i, A_i)$		
			$D1_j$		
$C3_i$	$B1_i$	A_i	True	False	Total
True	True	True	0.637681[a]	0.362319	1.000000
True	True	False	0.698713	0.301287	1.000000
True	False	True	0.747126	0.252874	1.000000
True	False	False	0.714509	0.285491	1.000000
False	True	True	0.727848	0.272152	1.000000
False	True	False	0.714589	0.285411	1.000000
False	False	True	0.749340	0.250660	1.000000
False	False	False	0.712410	0.287590	1.000000

[a]From Table 6.22, 0.637681 = 88/138

Step 9: Compute Posterior Probabilities

We compute the *posteriors* for $P(C3_j|D1_i, B1_i, A_i)$ and report this in Table 9.270.
We compute the *posteriors* for $P(B1_j|D1_i, C3_i, A_i)$ and report this in Table 9.271.
We compute the *posteriors* for $P(A_j|D1_i, C3_i, B1_i)$ and report this in Table 9.272.

9.6.4   Path 4: BBN Solution Protocol

Step 2: Specify the BBN–[$A \rightarrow B1|A \rightarrow C1|B1 \rightarrow C1|B1 \rightarrow D2|C1 \rightarrow D2$]

See Fig. 9.35.

Table 9.269 Joint and marginal probabilities

$P(D1_j, C3_i, B1_i, A_i)$

$D1_j$

$C3_i$	$B1_i$	A_i	True	False	Marginals
True	True	True	0.008967[a]	0.005095	0.014062
True	True	False	0.096128	0.041451	0.137579
True	False	True	0.025804	0.008734	0.034538
True	False	False	0.224586	0.089736	0.314321
False	True	True	0.011309	0.004229	0.015538
False	True	False	0.096985	0.038736	0.135721
False	False	True	0.028596	0.009566	0.038162
False	False	False	0.220903	0.089176	0.310079
Marginals			0.713278	0.286722	1.000000

Note: ZSMS = 0.000122 − 0.000122 = 0.000000
[a]From Tables 9.265, 9.266, 9.267, and 9.268, 0.008967 = 0.637681 * 0.475073 * 0.289345 * 0.102300

Table 9.270 Posterior probabilities

$P(C3_j | D1_i, B1_i, A_i)$

$C3_j$

$D1_i$	$B1_i$	A_i	True	False	Total
True	True	True	0.442248[a]	0.557752	1.000000
True	True	False	0.497781	0.502219	1.000000
True	False	True	0.474335	0.525665	1.000000
True	False	False	0.504133	0.495867	1.000000
False	True	True	0.546459	0.453541	1.000000
False	True	False	0.516925	0.483075	1.000000
False	False	True	0.477267	0.522733	1.000000
False	False	False	0.501566	0.498434	1.000000

[a]From Table 9.269, 0.442248 = 0.008967/(0.008967 + 0.011309)

Table 9.271 Posterior probabilities

$P(B1_j | D1_i, C3_i, A_i)$

$B1_j$

$D1_i$	$C3_i$	A_i	True	False	Total
True	True	True	0.257890[a]	0.742110	1.000000
True	True	False	0.299732	0.700268	1.000000
True	False	True	0.283398	0.716602	1.000000
True	False	False	0.305092	0.694908	1.000000
False	True	True	0.368436	0.631564	1.000000
False	True	False	0.315967	0.684033	1.000000
False	False	True	0.306549	0.693451	1.000000
False	False	False	0.302836	0.697164	1.000000

[a]From Table 9.269, 0.257890 = 0.008967/(0.008967 + 0.025804)

Table 9.272 Posterior probabilities

			$P(A_j\|D1_i, C3_i, B1_i)$		
			A_j		
$D1_i$	$C3_i$	$B1_i$	True	False	Total
True	True	True	0.085324[a]	0.914676	1.000000
True	True	False	0.103056	0.896944	1.000000
True	False	True	0.104430	0.895570	1.000000
True	False	False	0.114615	0.885385	1.000000
False	True	True	0.109462	0.890538	1.000000
False	True	False	0.088694	0.911306	1.000000
False	False	True	0.098421	0.901579	1.000000
False	False	False	0.096877	0.903123	1.000000

[a]From Table 9.269, 0.085324 = 0.008967/(0.008967 + 0.096128)

Fig. 9.35 Paths
$[A \rightarrow B1|A \rightarrow C1|B1 \rightarrow C1|B1 \rightarrow D2|C1 \rightarrow D2]$

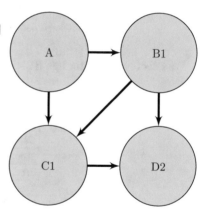

Step 6: Determine Prior or Unconditional Probabilities

We select Node A for this BBN structure as reported in Table 9.273.

Table 9.273 Node A: prior probabilities

$P(A_j)$		
True	False	Total
0.102300[a]	0.897700	1.000000

[a]From Table 6.5, 0.102300 = 1023/10000

Step 7: Determine Likelihood Probabilities

We select Node $B1$ for this BBN structure as reported in Table 9.274.
We select Node $C1$ for this BBN structure as reported in Table 9.275.

Table 9.274 Node $B1$: likelihood probabilities

$P(B1_j|A_i)$			
$B1_j$			
A_i	True	False	Total
True	0.289345[a]	0.710655	1.000000
False	0.304445	0.695555	1.000000

[a]From Table 6.8, 0.289345 = 296/1023

Table 9.275 Node $C1$: likelihood probabilities

$P(C1_j|B1_i, A_i)$				
$C1_j$				
$B1_i$	A_i	True	False	Total
True	True	0.466216[a]	0.533784	1.000000
True	False	0.483352	0.516648	1.000000
False	True	0.478680	0.521320	1.000000
False	False	0.512172	0.487828	1.000000

[a]From Table 6.13, 0.466216 = 138/296

We select Node $D2$ for this BBN structure as reported in Table 9.276.

Table 9.276 Node $D2$: likelihood probabilities

$P(D2_j|C1_i, B1_i, A_i)$					
$D2_j$					
$C1_i$	$B1_i$	A_i	True	False	Total
True	True	True	0.692940[a]	0.307060	1.000000
True	True	False	0.692940	0.307060	1.000000
True	False	True	0.717710	0.282290	1.000000
True	False	False	0.717710	0.282290	1.000000
False	True	True	0.715924	0.284076	1.000000
False	True	False	0.715924	0.284076	1.000000
False	False	True	0.716496	0.283504	1.000000
False	False	False	0.716496	0.283504	1.000000

[a]From Table 6.24, 0.692940 = 1011/1459

Step 8: Determine Joint and Marginal Probabilities

We compute the *joints* and *marginals* for $P(D2_j, C1_i, B1_i, A_i)$ and report these in Table 9.277.

Step 9: Compute Posterior Probabilities

We compute the *posteriors* for $P(C1_j|D2_i, B1_i, A_i)$ and report this in Table 9.278.

Table 9.277 Joint and marginal probabilities

| | | | $P(D2_j, C1_i, B1_i, A_i)$ | | |
| | | | $D2_j$ | | |
$C1_i$	$B1_i$	A_i	True	False	Marginals
True	True	True	0.009563[a]	0.004237	0.013800
True	True	False	0.091537	0.040563	0.132100
True	False	True	0.024976	0.009824	0.034800
True	False	False	0.229524	0.090276	0.319800
False	True	True	0.011312	0.004488	0.015800
False	True	False	0.101088	0.040112	0.141200
False	False	True	0.027155	0.010745	0.037900
False	False	False	0.218245	0.086355	0.304600
Marginals			0.713400	0.286600	1.000000

Note: ZSMS $= 0.000000 + 0.000000 = 0.000000$
[a]From Tables 9.273, 9.274, 9.275, and 9.276, $0.009563 = 0.692940 * 0.466216 * 0.289345 * 0.102300$

Table 9.278 Posterior probabilities

| | | | $P(C1_j | D2_i, B1_i, A_i)$ | | |
| | | | $C1_j$ | | |
$D2_i$	$B1_i$	A_i	True	False	Total
True	True	True	0.458106[a]	0.541894	1.000000
True	True	False	0.475208	0.524792	1.000000
True	False	True	0.479102	0.520898	1.000000
True	False	False	0.512595	0.487405	1.000000
False	True	True	0.485618	0.514382	1.000000
False	True	False	0.502795	0.497205	1.000000
False	False	True	0.477609	0.522391	1.000000
False	False	False	0.511100	0.488900	1.000000

[a]From Table 9.277, $0.458106 = 0.009563/(0.009563 + 0.011312)$

We compute the *posteriors* for $P(B1_j | D2_i, C1_i, A_i)$ and report this in Table 9.279.
We compute the *posteriors* for $P(A_j | D2_i, C1_i, B1_i)$ and report this in Table 9.280.

9.6.5 Path 5: BBN Solution Protocol

Step 2: Specify the BBN–$[A \rightarrow B1 | A \rightarrow D1 | B1 \rightarrow C2 | B1 \rightarrow D1 | C2 \rightarrow D1]$

See Fig. 9.36.

Table 9.279 Posterior probabilities

$P(B1_j | D2_i, C1_i, A_i)$

			$B1_j$		
$D2_i$	$C1_i$	A_i	True	False	Total
True	True	True	0.276864[a]	0.723136	1.000000
True	True	False	0.285109	0.714891	1.000000
True	False	True	0.294061	0.705939	1.000000
True	False	False	0.316561	0.683439	1.000000
False	True	True	0.301358	0.698642	1.000000
False	True	False	0.310019	0.689981	1.000000
False	False	True	0.294646	0.705354	1.000000
False	False	False	0.317171	0.682829	1.000000

[a]From Table 9.277, 0.276864 = 0.009563/(0.009563 + 0.024976)

Table 9.280 Posterior probabilities

$P(A_j | D2_i, C1_i, B1_i)$

			A_j		
$D2_i$	$C1_i$	$B1_i$	True	False	Total
True	True	True	0.094585[a]	0.905415	1.000000
True	True	False	0.098139	0.901861	1.000000
True	False	True	0.100637	0.899363	1.000000
True	False	False	0.110657	0.889343	1.000000
False	True	True	0.094585	0.905415	1.000000
False	True	False	0.098139	0.901861	1.000000
False	False	True	0.100637	0.899363	1.000000
False	False	False	0.110657	0.889343	1.000000

[a]From Table 9.277, 0.094585 = 0.009563/(0.009563 + 0.091537)

Step 6: Determine Prior or Unconditional Probabilities

We select Node A for this BBN structure as reported in Table 9.281.

Table 9.281 Node A: prior probabilities

$P(A_j)$

True	False	Total
0.102300[a]	0.897700	1.000000

[a]From Table 6.5, 0.102300 = 1023/10000

Step 7: Determine Likelihood Probabilities

We select Node $B1$ for this BBN structure as reported in Table 9.282.

Fig. 9.36 Paths
$[A \rightarrow B1 | A \rightarrow D1 | B1 \rightarrow C2 | B1 \rightarrow D1 | C2 \rightarrow D1]$

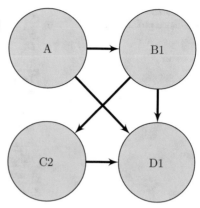

Table 9.282 Node $B1$:
likelihood probabilities

$P(B1_j | A_i)$			
$B1_j$			
A_i	True	False	Total
True	0.289345[a]	0.710655	1.000000
False	0.304445	0.695555	1.000000

[a]From Table 6.8, 0.289345 = 296/1023

We select Node $C2$ for this BBN structure as reported in Table 9.283.

Table 9.283 Node $C2$:
likelihood probabilities

$P(C2_j | B1_i, A_i)$				
$C2_j$				
$B1_i$	A_i	True	False	Total
True	True	0.481677[a]	0.518323	1.000000
True	False	0.481677	0.518323	1.000000
False	True	0.508679	0.491321	1.000000
False	False	0.508679	0.491321	1.000000

[a]From Table 6.15, 0.481677 = 1459/3029

We select Node $D1$ for this BBN structure as reported in Table 9.284.

Step 8: Determine Joint and Marginal Probabilities

We compute the *joints* and *marginals* for $P(D1_j, C2_i, B1_i, A_i)$ and report these in
Table 9.285.

Table 9.284 Node $D1$:
likelihood probabilities

$P(D1_j$	$|C2_i,$	$B1_i,$	$A_i)$			
$D1_j$						
$C2_i$	$B1_i$	A_i	True	False	Total	
True	True	True	0.637681[a]	0.362319	1.000000	
True	True	False	0.698713	0.301287	1.000000	
True	False	True	0.747126	0.252874	1.000000	
True	False	False	0.714509	0.285491	1.000000	
False	True	True	0.727848	0.272152	1.000000	
False	True	False	0.714589	0.285411	1.000000	
False	False	True	0.749340	0.250660	1.000000	
False	False	False	0.712410	0.287590	1.000000	

[a]From Table 6.22, $0.637681 = 88/138$

Table 9.285 Joint and
marginal probabilities

$P(D1_j,$	$C2_i,$	$B1_i,$	$A_i)$		
$D1_j$					
$C2_i$	$B1_i$	A_i	True	False	Marginals
True	True	True	0.009092[a]	0.005166	0.014258
True	True	False	0.091980	0.039662	0.131642
True	False	True	0.027629	0.009352	0.036981
True	False	False	0.226942	0.090677	0.317619
False	True	True	0.011167	0.004175	0.015342
False	True	False	0.101227	0.040431	0.141658
False	False	True	0.026766	0.008953	0.035719
False	False	False	0.218554	0.088227	0.306781
Marginals			0.713357	0.286643	1.000000

Note: ZSMS $= 0.000043 - 0.000043 = 0.000000$
[a]From Tables 9.281, 9.282, 9.283, and 9.284, $0.009092 = 0.637681 * 0.481677 * 0.289345 * 0.102300$

Step 9: Compute Posterior Probabilities

We compute the *posteriors* for $P(C2_j|D1_i, B1_i, A_i)$, and report this in Table 9.286.
We compute the *posteriors* for $P(B1_j|D1_i, C2_i, A_i)$ and report this in Table 9.287.
We compute the *posteriors* for $P(A_j|D1_i, C2_i, B1_i)$ and report this in Table 9.288.

9.6.6   Path 6: BBN Solution Protocol

Step 2: Specify the BBN–$[A \rightarrow C1|A \rightarrow D1|B2 \rightarrow C1|B2 \rightarrow D1|C1 \rightarrow D1]$

See Fig. 9.37.

Table 9.286 Posterior probabilities

			$P(C2_j \mid D1_i, B1_i, A_i)$		
			$C2_j$		
$D1_i$	$B1_i$	A_i	True	False	Total
True	True	True	0.448786[a]	0.551214	1.000000
True	True	False	0.476070	0.523930	1.000000
True	False	True	0.507939	0.492061	1.000000
True	False	False	0.509414	0.490586	1.000000
False	True	True	0.553010	0.446990	1.000000
False	True	False	0.495202	0.504798	1.000000
False	False	True	0.510876	0.489124	1.000000
False	False	False	0.506848	0.493152	1.000000

[a]From Table 9.285, 0.448786 = 0.009092/(0.009092 + 0.011167)

Table 9.287 Posterior probabilities

			$P(B1_j \mid D1_i, C2_i, A_i)$		
			$B1_j$		
$D1_i$	$C2_i$	A_i	True	False	Total
True	True	True	0.247590[a]	0.752410	1.000000
True	True	False	0.288410	0.711590	1.000000
True	False	True	0.294388	0.705612	1.000000
True	False	False	0.316551	0.683449	1.000000
False	True	True	0.355838	0.644162	1.000000
False	True	False	0.304299	0.695701	1.000000
False	False	True	0.318038	0.681962	1.000000
False	False	False	0.314249	0.685751	1.000000

[a]From Table 9.285, 0.247590 = 0.009092/(0.009092 + 0.027629)

Table 9.288 Posterior probabilities

			$P(A_j \mid D1_i, C2_i, B1_i)$		
			A_j		
$D1_i$	$C2_i$	$B1_i$	True	False	Total
True	True	True	0.089954[a]	0.910046	1.000000
True	True	False	0.108533	0.891467	1.000000
True	False	True	0.099355	0.900645	1.000000
True	False	False	0.109106	0.890894	1.000000
False	True	True	0.115236	0.884764	1.000000
False	True	False	0.093488	0.906512	1.000000
False	False	True	0.093607	0.906393	1.000000
False	False	False	0.092131	0.907869	1.000000

[a]From Table 9.285, 0.089954 = 0.009092/(0.009092 + 0.091980)

Fig. 9.37 Paths
$[A \rightarrow C1|A \rightarrow D1|B2 \rightarrow C1|B2 \rightarrow D1|C1 \rightarrow D1]$

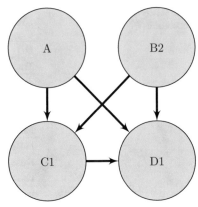

Step 6: Determine Prior or Unconditional Probabilities

We select Node *A* for this BBN structure as reported in Table 9.289.

Table 9.289 Node *A*: prior
probabilities

$P(A_j)$		
True	False	Total
0.102300[a]	0.897700	1.000000

[a]From Table 6.5, 0.102300 = 1023/10000

Step 7: Determine Likelihood Probabilities

We select Node *B2* for this BBN structure as reported in Table 9.290.

Table 9.290 Node *B2*:
likelihood probabilities

$P(B2_j|A_i)$			
$B2_j$			
A_i	True	False	Total
True	0.302900[a]	0.697100	1.000000
False	0.302900	0.697100	1.000000

[a]From Table 6.10, 0.302900 = 3029/10000

We select Node *C1* for this BBN structure as reported in Table 9.291.
We select Node *D1* for this BBN structure as reported in Table 9.292.

Table 9.291 Node $C1$:
likelihood probabilities

$P(C1_j\|B2_i, A_i)$					
$C1_j$					
$B2_i$	A_i	True	False	Total	
True	True	0.466216[a]	0.533784	1.000000	
True	False	0.483352	0.516648	1.000000	
False	True	0.478680	0.521320	1.000000	
False	False	0.512172	0.487828	1.000000	

[a]From Table 6.13, $0.466216 = 138/296$

Table 9.292 Node $D1$:
likelihood probabilities

$P(D1_j\|C1_i, B2_i, A_i)$					
$D1_j$					
$C1_i$	$B2_i$	A_i	True	False	Total
True	True	True	0.637681[a]	0.362319	1.000000
True	True	False	0.698713	0.301287	1.000000
True	False	True	0.747126	0.252874	1.000000
True	False	False	0.714509	0.285491	1.000000
False	True	True	0.727848	0.272152	1.000000
False	True	False	0.714589	0.285411	1.000000
False	False	True	0.749340	0.250660	1.000000
False	False	False	0.712410	0.287590	1.000000

[a]From Table 6.22, $0.637681 = 88/138$

Step 8: Determine Joint and Marginal Probabilities

We compute the *joints* and *marginals* for $P(D1_j, C1_i, B2_i, A_i)$ and report these in
Table 9.293.

Table 9.293 Joint and
marginal probabilities

$P(D1_j, C1_i, B2_i, A_i)$					
$D1_j$					
$C1_i$	$B2_i$	A_i	True	False	Marginals
True	True	True	0.009212[a]	0.005234	0.014446
True	True	False	0.091832	0.039598	0.131430
True	False	True	0.025504	0.008632	0.034136
True	False	False	0.229007	0.091503	0.320510
False	True	True	0.012039	0.004501	0.016540
False	True	False	0.100388	0.040096	0.140484
False	False	True	0.027858	0.009319	0.037177
False	False	False	0.217482	0.087795	0.305276
Marginals			0.713322	0.286678	1.000000

Note: ZSMS $= 0.000078 - 0.000078 = 0.000000$
[a]From Tables 9.289, 9.290, 9.291, and 9.292, $0.009212 =$
$0.637681 * 0.466216 * 0.302900 * 0.102300$

Step 9: Compute Posterior Probabilities

We compute the *posteriors* for $P(C1_j|D1_i, B2_i, A_i)$ and report this in Table 9.294.

Table 9.294 Posterior probabilities

$P(C1_j|D1_i, B2_i, A_i)$					
$C1_j$					
$D1_i$	$B2_i$	A_i	True	False	Total
True	True	True	0.433498[a]	0.566502	1.000000
True	True	False	0.477743	0.522257	1.000000
True	False	True	0.477941	0.522059	1.000000
True	False	False	0.512907	0.487093	1.000000
False	True	True	0.537634	0.462366	1.000000
False	True	False	0.496879	0.503121	1.000000
False	False	True	0.480874	0.519126	1.000000
False	False	False	0.510341	0.489659	1.000000

[a]From Table 9.293, $0.433498 = 0.009212/(0.009212 + 0.012039)$

We compute the *posteriors* for $P(B2_j|D1_i, C1_i, A_i)$ and report this in Table 9.295.

Table 9.295 Posterior probabilities

$P(B2_j|D1_i, C1_i, A_i)$					
$B2_j$					
$D1_i$	$C1_i$	A_i	True	False	Total
True	True	True	0.265358[a]	0.734642	1.000000
True	True	False	0.286223	0.713777	1.000000
True	False	True	0.301745	0.698255	1.000000
True	False	False	0.315815	0.684185	1.000000
False	True	True	0.377477	0.622523	1.000000
False	True	False	0.302043	0.697957	1.000000
False	False	True	0.325714	0.674286	1.000000
False	False	False	0.313516	0.686484	1.000000

[a]From Table 9.293, $0.265358 = 0.009212/(0.009212 + 0.025504)$

We compute the *posteriors* for $P(A_j|D1_i, C1_i, B2_i)$ and report this in Table 9.296.

Table 9.296 Posterior probabilities

$P(A_j|D1_i, C1_i, B2_i)$					
A_j					
$D1_i$	$C1_i$	$B2_i$	True	False	Total
True	True	True	0.091171[a]	0.908829	1.000000
True	True	False	0.100208	0.899792	1.000000
True	False	True	0.107081	0.892919	1.000000
True	False	False	0.113550	0.886450	1.000000
False	True	True	0.116751	0.883249	1.000000
False	True	False	0.086205	0.913795	1.000000
False	False	True	0.100936	0.899064	1.000000
False	False	False	0.095958	0.904042	1.000000

[a]From Table 9.293, 0.091171 = 0.009212/(0.009212 + 0.091832)

Fig. 9.38 Paths
$[A \rightarrow B1|A \rightarrow C1|A \rightarrow D1|$
$B1 \rightarrow C1|B1 \rightarrow D1|C1 \rightarrow D1]$

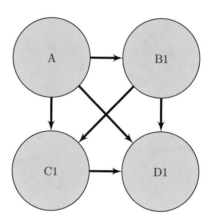

9.7   6-Path BBN

9.7.1   Path 1: BBN Solution Protocol

Step 2: Specify the
BBN–$[A \rightarrow B1|A \rightarrow C1|A \rightarrow D1|B1 \rightarrow C1|B1 \rightarrow D1|C1 \rightarrow D1]$

See Fig. 9.38.

Step 6: Determine Prior or Unconditional Probabilities

We select Node A for this BBN structure as reported in Table 9.297.

Table 9.297 Node A: prior probabilities

$P(A_j)$		
True	False	Total
0.102300[a]	0.897700	1.000000

[a]From Table 6.5, $0.102300 = 1023/10000$

Step 7: Determine Likelihood Probabilities

We select Node $B1$ for this BBN structure as reported in Table 9.298.

Table 9.298 Node $B1$: likelihood probabilities

$P(B1_j\|A_i)$			
$B1_j$			
A_i	True	False	Total
True	0.289345[a]	0.710655	1.000000
False	0.304445	0.695555	1.000000

[a]From Table 6.8, $0.289345 = 296/1023$

We select Node $C1$ for this BBN structure as reported in Table 9.299.

Table 9.299 Node $C1$: likelihood probabilities

$P(C1_j\|B1_i, A_i)$				
$C1_j$				
$B1_i$	A_i	True	False	Total
True	True	0.466216[a]	0.533784	1.000000
True	False	0.483352	0.516648	1.000000
False	True	0.478680	0.521320	1.000000
False	False	0.512172	0.487828	1.000000

[a]From Table 6.13, $0.466216 = 138/296$

We select Node $D1$ for this BBN structure as reported in Table 9.300.

Step 8: Determine Joint and Marginal Probabilities

We compute the *joints* and *marginals* for $P(D1_j, C1_i, B1_i, A_i)$ and report these in Table 9.301.

Step 9: Compute Posterior Probabilities

We compute the *posteriors* for $P(C1_j|D1_i, B1_i, A_i)$ and report this in Table 9.302.
We compute the *posteriors* for $P(B1_j|D1_i, C1_i, A_i)$ and report this in Table 9.303.
We compute the *posteriors* for $P(A_j|D1_i, C1_i, B1_i)$ and report this in Table 9.304.

Table 9.300 Node $D1$: likelihood probabilities

$P(D1_j\|C1_i, B1_i, A_i)$					
$D1_j$					
$C1_i$	$B1_i$	A_i	True	False	Total
True	True	True	0.637681[a]	0.362319	1.000000
True	True	False	0.698713	0.301287	1.000000
True	False	True	0.747126	0.252874	1.000000
True	False	False	0.714509	0.285491	1.000000
False	True	True	0.727848	0.272152	1.000000
False	True	False	0.714589	0.285411	1.000000
False	False	True	0.749340	0.250660	1.000000
False	False	False	0.712410	0.287590	1.000000

[a]From Table 6.22, 0.637681 = 88/138

Table 9.301 Joint and marginall probabilities

$P(D1_j, C1_i, B1_i, A_i)$					
$D1_j$					
$C1_i$	$B1_i$	A_i	True	False	Marginals
True	True	True	0.008800[a]	0.005000	0.013800
True	True	False	0.092300	0.039800	0.132100
True	False	True	0.026000	0.008800	0.034800
True	False	False	0.228500	0.091300	0.319800
False	True	True	0.011500	0.004300	0.015800
False	True	False	0.100900	0.040300	0.141200
False	False	True	0.028400	0.009500	0.037900
False	False	False	0.217000	0.087600	0.304600
Marginals			0.713400	0.286600	1.000000

Note: ZSMS = 0.000078 − 0.000078 = 0.000000
[a]From Tables 9.297, 9.298, 9.299, and 9.300, 0.008800 = 0.637681 * 0.466216 * 0.289345 * 0.102300

Table 9.302 Posterior probabilities

$P(C1_j\|D1_i, B1_i, A_i)$					
$C1_j$					
$D1_i$	$B1_i$	A_i	True	False	Total
True	True	True	0.433498[a]	0.566502	1.000000
True	True	False	0.477743	0.522257	1.000000
True	False	True	0.477941	0.522059	1.000000
True	False	False	0.512907	0.487093	1.000000
False	True	True	0.537634	0.462366	1.000000
False	True	False	0.496879	0.503121	1.000000
False	False	True	0.480874	0.519126	1.000000
False	False	False	0.510341	0.489659	1.000000

[a]From Table 9.301, 0.433498 = 0.008800/(0.008800 + 0.011500)

Table 9.303 Posterior probabilities

$P(B1_j \mid D1_i, C1_i, A_i)$					
$B1_j$					
$D1_i$	$C1_i$	A_i	True	False	Total
True	True	True	0.252874[a]	0.747126	1.000000
True	True	False	0.287718	0.712282	1.000000
True	False	True	0.288221	0.711779	1.000000
True	False	False	0.317395	0.682605	1.000000
False	True	True	0.362319	0.637681	1.000000
False	True	False	0.303585	0.696415	1.000000
False	False	True	0.311594	0.688406	1.000000
False	False	False	0.315090	0.684910	1.000000

[a]From Table 9.301, $0.252874 = 0.008800/(0.008800 + 0.026000)$

Table 9.304 Posterior probabilities

$P(A_j \mid D1_i, C1_i, B1_i)$					
A_j					
$D1_i$	$C1_i$	$B1_i$	True	False	Total
True	True	True	0.087043[a]	0.912957	1.000000
True	True	False	0.102161	0.897839	1.000000
True	False	True	0.102313	0.897687	1.000000
True	False	False	0.115729	0.884271	1.000000
False	True	True	0.111607	0.888393	1.000000
False	True	False	0.087912	0.912088	1.000000
False	False	True	0.096413	0.903587	1.000000
False	False	False	0.097837	0.902163	1.000000

[a]From Table 9.301, $0.087043 = 0.008800/(0.008800 + 0.092300)$